16th Edition IEE Wiring Regulations

Explained and Illustrated

By the same author

Electrical Installation Work

Wiring Regulations: Inspection, Testing and Certification

Wiring Regulations: Design and Verification

Wiring Systems and Fault Finding for Installation Electricians

Electric Wiring Domestic

PAT: Portable Appliance Testing

16th Edition IEE Wiring Regulations
Explained and Illustrated

Seventh Edition

Brian Scaddan IEng, MIIE (elec)

AMSTERDAM • BOSTON • HEIDELBERG • LONDON • NEW YORK • OXFORD
PARIS • SAN DIEGO • SAN FRANCISCO • SINGAPORE • SYDNEY • TOKYO

Newnes is an imprint of Elsevier

ELSEVIER

Newnes

Newnes
An imprint of Elsevier
Linacre House, Jordan Hill, Oxford OX2 8DP
30 Corporate Drive, Burlington, MA 01803

First published 1989
Second edition 1991
Third edition 1996
Fourth edition 1998
Fifth edition 2001
Sixth edition 2002
Reprinted 2002, 2003, 2004
Seventh edition 2005
Reprinted 2005

British Library Cataloguing in Publication Data
A catalogue record for this book is available from the British Library

Library of Congress Cataloguing in Publication Data
A catalogue record for this book is available from the Library of Congress

ISBN 0 7506 6539 4

For information on all Newnes publications
visit our website at www.newnespress.com

Working together to grow
libraries in developing countries

www.elsevier.com | www.bookaid.org | www.sabre.org

ELSEVIER   BOOK AID
            International   Sabre Foundation

Typeset by Integra Software Services Pvt. Ltd, Pondicherry, India
www.integra-india.com

Printed and bound in Great Britain by Biddles Ltd, King's Lynn, Norfolk

CONTENTS

Contents

PREFACE

As a result of many years developing and teaching courses devoted to compliance with the IEE Wiring Regulations, it has become apparent to me that many operatives and personnel in the electrical contracting industry have forgotten the basic principles and concepts upon which electric power supply and its use are based. As a result of this, misconceived ideas and much confusion have arisen over the interpretation of the Regulations.

It is the intention of this book to dispel such misconceptions and to educate and where necessary refresh the memory of the reader. In this respect, emphasis has been placed on those areas where most confusion arises, namely earthing and bonding, protection, and circuit design. Much of Part 5 of the Regulations is not mentioned, since it deals with selection of accessories etc. which needs little or no explanation. Part 6 which deals with special installations or locations, has now been included.

The current sixteenth edition of the IEE Wiring Regulations, also known as BS 7671, to which this book conforms, was published in June 2001. The philosophy and concepts that this book seeks to explain remain unchanged, regardless of the edition. It is *not* a guide to the Regulations or a replacement for them; nor does it seek to interpret them Regulation by Regulation. It should, in fact, be read in conjunction with them; to help the reader, each chapter cites the relevant Regulation numbers for cross-reference.

It is hoped that the book will be found particularly useful by college students, electricians and technicians, and also by managers of smaller electrical contracting firms that do no normally employ engineers or designers. It should also be a useful addition to the library of those studying for the C & G 2381 qualification.

INTRODUCTION

It was once said, by whom I have no idea, that 'rules and regulations are for the guidance of wise men and the blind obedience of fools.' This is certainly true in the case of the IEE Wiring (BS 7671) Regulations. They are not statutory rules, but recommendations for the safe selection and erection of wiring installations. Earlier editions were treated as an 'electrician's Bible': the Regulations now take the form primarily of a design document.

The IEE Wiring Regulations are divided into seven parts. These follow a logical pattern from the basic requirements to the final testing and inspection of an installation:

Part 1 indicates the range and type of installations covered by the Regulations, what they are intended for, and the basic requirements for safety.

Part 2 is devoted to the definitions of the terms used throughout the Regulations.

Part 3 details the general information needed before any design work can usefully proceed.

Part 4 informs the designer of the different methods available for protection against electric shock, overcurrent etc., and how to apply those methods.

Part 5 enables the correct type of equipment, cable, accessory etc. to be selected in accordance with the requirements of Parts 1–4.

Part 6 deals with particular requirements for special installations such as bathrooms, swimming pools, construction sites etc.

Part 7 provides details of the relevant tests to be performed on a completed installation before it is energized.

Appendices 1–7 provide tabulated and other background information required by the designer.

It must be remembered that the Regulations are not a collection of unrelated statements each to be interpreted in isolation; there are many cross-references throughout which may render such an interpretation valueless.

In using the Regulations I have found the index an invaluable starting place when seeking information. However, one may have to try different combinations of wording in order to locate a particular item. For example, determining how often an RCD should be tested via its test button could prove difficult since no reference is made under 'Residual current devices' or 'Testing'; however, 'Periodic testing' leads to Regulation 514–12, and the information in question is found in 514–12–02. In the index, this Regulation is referred to under 'Notices'.

1
FUNDAMENTAL REQUIREMENTS FOR SAFETY

(IEE Regs Part 1 and Chapter 13)

It does not require a degree in electrical engineering to realize that electricity at *low* voltage can, if uncontrolled, present a serious threat of injury to persons or livestock, or damage to property by fire.

Clearly the type and arrangement of the equipment used, together with the quality of workmanship provided, will go a long way to minimizing danger. The following is a list of basic requirements:

1 Use good workmanship.
2 Use approved materials and equipment.
3 Ensure that the correct type, size and current-carrying capacity of cables are chosen.
4 Ensure that equipment is suitable for the maximum power demanded of it.
5 Make sure that conductors are insulated, and sheathed or protected if necessary, or are placed in a position to prevent danger.
6 Joints and connections should be properly constructed to be mechanically and electrically sound.
7 Always provide overcurrent protection for every circuit in an installation (the protection for the whole installation is usually

provided by the supply authority), and ensure that protective devices are suitably chosen for their location and the duty they have to perform.

8 Where there is a chance of metalwork becoming live owing to a fault, it should be earthed, and the circuit concerned should be protected by an overcurrent device or a residual current device (RCD).

9 Ensure that all necessary bonding of services is carried out.

10 Do not place a fuse, a switch or a circuit breaker, unless it is a linked switch or circuit breaker, in an earthed neutral conductor. The linked type must be arranged to break all the phase conductors.

11 All single-pole switches must be wired in the phase conductor only.

12 A readily accessible and effective means of isolation must be provided, so that all voltage may be cut off from an installation or any of its circuits.

13 All motors must have a readily accessible means of disconnection.

14 Ensure that any item of equipment which may normally need operating or attending to by persons is accessible and easily operated.

15 Any equipment required to be installed in a situation exposed to weather or corrosion, or in explosive or volatile environments, should be of the correct type for such adverse conditions.

16 Before adding to or altering an installation, ensure that such work will not impair any part of the existing installation.

17 After completion of an installation or an alteration to an installation, the work must be inspected and tested to ensure, as far as reasonably practicable, that the fundamental requirements for safety have been met.

These requirements form the basis of the IEE Regulations.

It is interesting to note that, whilst the Wiring Regulations are not statutory, they may be used to claim compliance with Statutory Regulations such as the Electricity at Work Regulations, and the

Health and Safety at Work Act. In fact, the Health and Safety Executive produces guidance notes for installations in such places as schools and construction sites. The contents of these documents reinforce and extend the requirements of the IEE Regulations. Extracts from the Health and Safety at Work Act and the Electricity at Work Regulations are reproduced below.

The Health and Safety at Work Act 1974

Duties of employers

Employers must safeguard, as far as is reasonably practicable, the health, safety and welfare of all the people who work for them. This applies in particular to the provision and maintenance of safe plant and systems of work, and covers all machinery, equipment and appliances used.

Some examples of the matters which many employers need to consider are:

1 Is all plant up to the necessary standards with respect to safety and risk to health?
2 When new plant is installed, is latest good practice taken into account?
3 Are systems of work safe? Thorough checks of all operations, especially those operations carried out infrequently, will ensure that danger of injury or to health is minimized. This may require special safety systems, such as 'permits to work'.
4 Is the work environment regularly monitored to ensure that, where known toxic contaminants are present, protection conforms to current hygiene standards?
5 Is monitoring also carried out to check the adequacy of control measures?
6 Is safety equipment regularly inspected? All equipment and appliances for safety and health, such as personal protective equipment, dust and fume extraction, guards, safe access arrangement, monitoring and testing devices, need regular inspection (Section 2(1) and 2(2) of the Act).

No charge may be levied on any employee for anything done or provided to meet any specific requirement for health and safety at work (Section 9).

Risks to health from the use, storage, or transport of 'articles' and 'substances' must be minimized. The term *substance* is defined as 'any natural or artificial substance whether in solid or liquid form or in the form of gas or vapour' (Section 53(1)).

To meet these aims, all reasonably practicable precautions must be taken in the handling of any substance likely to cause a risk to health. Expert advice can be sought on the correct labelling of substances, and the suitability of containers and handling devices. All storage and transport arrangements should be kept under review.

Safety information and training

It is now the duty of employers to provide any necessary information and training in safe practices, including information on legal requirements.

Duties to others

Employers must also have regard for the health and safety of the self-employed or contractors' employees who may be working close to their own employees; and for the health and safety of the public who may be affected by their firm's activities.

Similar responsibilities apply to self-employed persons, manufacturers and suppliers.

Duties of employees

Employees have a duty under the Act to take reasonable care to avoid injury to themselves or to others by their work activities, and to cooperate with employers and others in meeting statutory requirements. The Act also requires employees not to interfere with or misuse anything provided to protect their health, safety or welfare in compliance with the Act.

The Electricity at Work Regulations 1989

Persons on whom duties are imposed by these Regulations

(1) Except where otherwise expressly provided in these Regulations, it shall be the duty of every:

    (a) employer and self-employed person to comply with the provisions of these Regulations in so far as they relate to matters which are within his control; and

    (b) manager of a mine or quarry (within in either case the meaning of Section 180 of the Mines and Quarries Act 1954*) to ensure that all requirements or prohibitions imposed by or under these Regulations are complied with in so far as they relate to the mine or quarry or part of a quarry of which he is the manager and to matters which are within his control.

(2) It shall be the duty of every employee while at work:

    (a) to cooperate with his employer in so far as is necessary to enable any duty placed on that employer by the provisions of these Regulations to be complied with; and

    (b) to comply with the provisions of these Regulations in so far as they relate to matters which are within his control.

Employer

(1) For the purposes of the Regulations, an employer is any person or body who (a) employs one or more individuals under a contract of employment or apprenticeship; or (b) provides training under the schemes to which the HSW Act applies through the Health and Safety (Training for Employment) Regulations 1988 (Statutory Instrument No. 1988/1222).

---

*   1954 C.70; Section 180 was amended by SI 1974/2013.

Self-employed

(2) A self-employed person is an individual who works for gain or reward otherwise than under a contract of employment whether or not he employs others.

Employee

(3) Regulation 3(2)(a) reiterates the duty placed on employees by Section 7(b) of the HSW Act.

(4) Regulation 3(2)(b) places duties on employees equivalent to those placed on employers and self-employed persons where these are matters within their control. This will include those trainees who will be considered as employees under the Regulations described in paragraph 1.

(5) This arrangement recognizes the level of responsibility which many employees in the electrical trades and professions are expected to take on as part of their job. The 'control' which they exercise over the electrical safety in any particular circumstances will determine to what extent they hold responsibilities under the Regulations to ensure that the Regulations are complied with.

(6) A person may find himself responsible for causing danger to arise elsewhere in an electrical system, at a point beyond his own installation. This situation may arise, for example, due to unauthorized or unscheduled back feeding from his installation onto the system, or to raising the fault power level on the system above rated and agreed maximum levels due to connecting extra generation capacity etc. Because such circumstances are 'within his control', the effect of Regulation 3 is to bring responsibilities for compliance with the rest of the regulations to that person, thus making him a duty holder.

Absolute/reasonably practicable

(7) Duties in some of the Regulations are subject to the qualifying term 'reasonably practicable'. Where qualifying terms are absent the requirement in the Regulation is said to be absolute. The meaning

of reasonably practicable has been well established in law. The interpretations below are given only as a guide to duty holders.

Absolute

(8) If the requirement in a Regulation is 'absolute', for example, if the requirement is not qualified by the words 'so far as is reasonably practicable', the requirement must be met regardless of cost or any other consideration. Certain of the regulations making such absolute requirements are subject to the Defence provision of Regulation 29.

Reasonably practicable

(9) Someone who is required to do something 'so far as is reasonably practicable' must assess, on the one hand, the magnitude of the risks of a particular work activity or environment and, on the other hand, the costs in terms of the physical difficulty, time, trouble and expense which would be involved in taking steps to eliminate or minimize those risks. If, for example, the risks to health and safety of a particular work process are very low, and the cost or technical difficulties of taking certain steps to prevent those risks are very high, it might not be reasonably practicable to take those steps. The greater the degree of risk, the less weight that can be given to the cost of measures needed to prevent that risk.

(10) In the context of the Regulations, where the risk is very often that of death, for example, from electrocution, and where the nature of the precautions which can be taken are so often very simple and cheap, e.g. insulation, the level of duty to prevent that danger approaches that of an absolute duty.

(11) The comparison does not include the financial standing of the duty holder. Furthermore, where someone is prosecuted for failing to comply with a duty 'so far as is reasonably practicable', it would be for the accused to show the court that it was not reasonably practicable for him to do more than he had in fact done to comply with the duty (Section 40 of the HSW Act).

Appendix 2 of the IEE Regulations lists all of the other Statutory Regulations and Memoranda with which electrical installations must comply.

It is interesting to note that if an installation fails to comply with Chapter 13 of the Regulations, the Supply Authority has the right to refuse to give a supply or, in certain circumstances, to disconnect it.

While we are on the subject of Supply Authorities, let us look at the current declared supply voltages and tolerances. In order to align with European Harmonized Standards, our historic 415 V/240 V declared supply voltages have now become 400 V/230 V. However, this is only a paper exercise, and it is unlikely that consumers will notice any difference for many years, if at all. Let me explain, using single phase as the example.

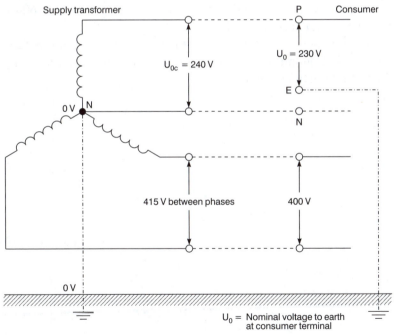

Note: The connection of the transformer star or neutral point to earth, helps to maintain that point at or very near zero volts.

Figure 1

The supply industry declared voltage was 240 V ± 6%, giving a range between 225.6 V and 254.4 V. The new values are 230 V + 10% − 6%, giving a range between 216.2 V and 253 V. Not a lot of difference. The industry has done nothing physical to reduce voltages from 240 V to 230 V, it is just the declaration that has been altered. Hence a measurement of voltage at supply terminals will give similar readings to those we have always known. In fact the presumed open circuit voltage, U_{oc}, at the supply transformer terminals is still 240 V.

Figure 1 shows the UK supply system and associated declared voltages.

BS 7671 details 2 voltage categories, Band 1 and Band 2. Band 1 is essentially Extra low voltage systems and Band 2, Low voltage systems.

ELV is less than 50 V ac between conductors or to earth. LV exceeds ELV up to 1000 V ac between conductors and 600 V between conductors and earth.

The suppliers are now governed by the "Electricity Safety, Quality & Continuity Regulations 2002" (formerly the Electricity Supply Regulations 1988).

2
EARTHING

(Relevant chapters and parts
Chapters 31, 33, 41, 47, 53, 54, 55, Part 6)

Definitions used in this chapter

Bonding conductor   A protective conductor providing equipotential bonding.

Circuit protective conductor   A protective conductor connecting exposed conductive parts of equipment to the main earthing terminal.

Direct contact   Contact of persons or livestock with live parts.

Earth   The conductive mass of earth, whose electric potential at any point is conventionally taken as zero.

Earth electrode resistance   The resistance of an earth electrode to earth.

Earth fault loop impedance   The impedance of the phase-to-earth loop path starting and ending at the point of fault.

Earthing conductor   A protective conductor connecting a main earthing terminal of an installation to an earth electrode or other means of earthing.

Equipotential bonding Electrical connection maintaining various exposed conductive parts and extraneous conductive parts at a substantially equal potential.

Exposed conductive part A conductive part of equipment which can be touched and which is not a live part but which may become live under fault conditions.

Extraneous conductive part A conductive part liable to introduce a potential, generally earth potential and not forming part of the electrical installation.

Functional earthing Connection to earth necessary for proper functioning of electrical equipment.

Indirect contact Contact of persons or livestock with exposed conductive parts made live by a fault.

Leakage current Electric current in an unwanted conductive part under normal operating conditions.

Live part A conductor or conductive part intended to be energized in normal use, including a neutral conductor but, by convention, not a PEN conductor.

PEN conductor A conductor combining the functions of both protective conductor and neutral conductor.

Phase conductor A conductor of an AC system for the transmission of electrical energy, other than a neutral conductor.

PME An earthing arrangement, found in TN– C–S systems, where an installation is earthed via the supply neutral conductor.

Protective conductor A conductor used for some measure of protection against electric shock and intended for connecting together any of the following parts:

exposed conductive parts
extraneous conductive parts
main earthing terminal
earth electrode(s)
earthed point of the source.

Residual current device   An electromechanical switching device or association of devices intended to cause the opening of the contacts when the residual current attains a given value under given conditions.

Simultaneously accessible parts   Conductors or conductive parts which can be touched simultaneously by a person or, where applicable, by livestock.

Earth: what it is, and why and how we connect to it

(IEE Regs Section 413)

The thin layer of material which covers our planet, be it rock, clay, chalk or whatever, is what we in the world of electricity refer to as earth. So, why do we need to connect anything to it? After all, it is not as if earth is a good conductor.

Perhaps it would be wise at this stage to investigate potential difference (PD). A potential difference is exactly what it says it is: a difference in potential (volts). Hence, two conductors having PDs of, say, 20 V and 26 V have a PD between them of $26 \backsim 20 = 6$ V. The original PDs, i.e. 20 V and 26 V, are the PDs between 20 V and 0 V and 26 V and 0 V.

So where does this 0 V or zero potential come from? The simple answer is, in our case, the earth. The definition of earth is therefore the conductive mass of earth, whose electric potential at any point is conventionally taken as zero.

Hence, if we connect a voltmeter between a live part (e.g. the phase conductor of, say, a socket outlet) and earth, we may read 230 V; the conductor is at 230 V, the earth at zero. The earth provides a path to complete the circuit. We would measure nothing at all if we connected our voltmeter between, say, the positive 12 V terminal of a car battery and earth, as in this case the earth plays no part in any circuit. Figure 2 illustrates this difference.

Hence, a person in an installation touching a live part whilst standing on the earth would take the place of the voltmeter in Figure 2(a), and could suffer a severe electric shock. Remember that the accepted lethal level of shock current passing through a

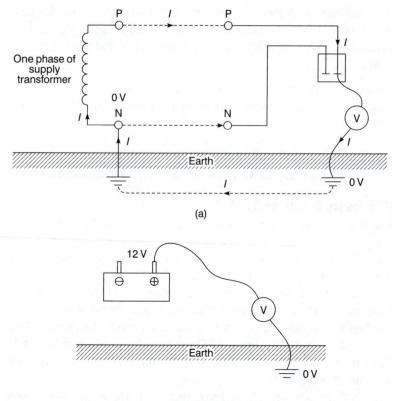

Figure 2

person is only 50 mA or 1/20 A. The same situation would arise if the person were touching say, a faulty appliance and a gas or water pipe (Figure 3).

One method of providing some measure of protection against these effects is to join together (bond) all metallic parts and connect them to earth. This ensures that all metalwork in a healthy situation is at or near zero volts, and under fault conditions all metalwork will rise to a similar potential. So, simultaneous contact with two such metal parts would not result in a dangerous shock, as there will be no significant PD between them. This method is known as earthed equipotential bonding.

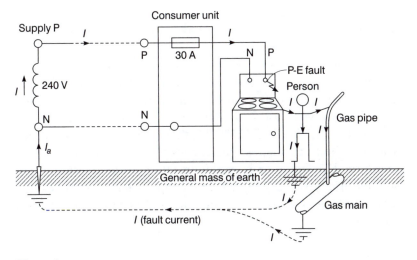

Figure 3

Unfortunately, as previously mentioned, earth itself is not a good conductor unless it is very wet, and therefore it presents a high resistance to the flow of fault current. This resistance is usually enough to restrict fault current to a level well below that of the rating of the protective device, leaving a faulty circuit uninterrupted. Clearly this is an unhealthy situation. The methods of overcoming this problem will be dealt with later.

In all but the most rural areas, consumers can connect to a metallic earth return conductor which is ultimately connected to the earthed neutral of the supply. This, of course, presents a low resistance path for fault currents to operate the protection.

Summarizing, then, connecting metalwork to earth, places that metal at or near zero potential, and bonding between metallic parts puts such parts at a similar potential even under fault conditions.

Connecting to earth

In the light of previous comments, it is obviously necessary to have as low an earth path resistance as possible, and the point of connection to earth is one place where such resistance may be reduced. When two conducting surfaces are placed in contact with each other, there will be a resistance to the flow of current dependent on the surface areas in contact. It is clear, then, that the greater surface contact area with earth that can be achieved, the better.

There are several methods of making a connection to earth, including the use of rods, plates and tapes. By far the most popular method in everyday use is the rod earth electrode. The plate type needs to be buried at a sufficient depth to be effective and, as such plates may be 1 or 2 metres square, considerable excavation may be necessary. The tape type is predominantly used in the earthing of large electricity substations, where the tape is laid in trenches in a mesh formation over the whole site. Items of plant are then earthed to this mesh.

Rod electrodes

These are usually of solid copper or copper-clad carbon steel, the latter being used for the larger-diameter rods with extension facilities. These facilities comprise: a thread at each end of the rod to enable a coupler to be used for connection of the next rod; a steel cap to protect the thread from damage when the rod is being driven in; a steel driving tip; and a clamp for the connection of an earth tape or conductor (Figure 4).

The choice of length and diameter of such a rod will, as previously mentioned, depend on the soil conditions. For example, a long thick electrode is used for earth with little moisture retention. Generally, a 1–2 m rod, 16 mm in diameter, will give a relatively low resistance.

Earth electrode resistance

(IEE Regs Definitions)

If we were to place an electrode in the earth and then measure the resistance between the electrode and points at increasingly larger distance from it, we would notice that the resistance increased with distance until a point was reached (usually around 2.5 m) beyond which no increase in resistance was noticed (Figure 5).

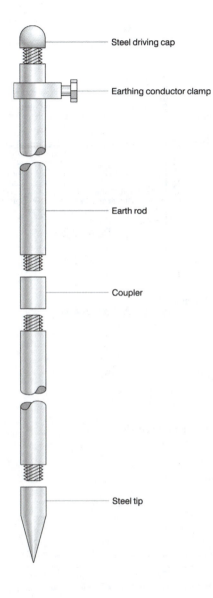

Figure 4

Earthing

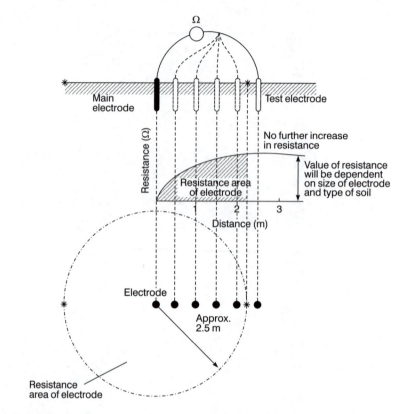

Figure 5   Resistance area of electrode

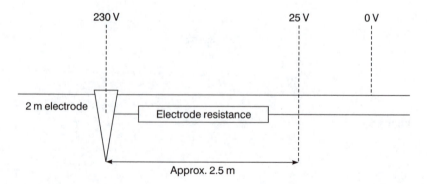

Figure 6

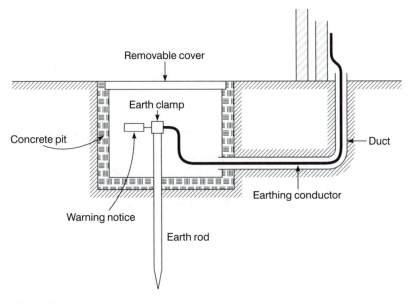

Figure 7

The resistance area around the electrode is particularly important with regard to the voltage at the surface of the ground (Figure 6). For a 2 m rod, with its top at ground level, 80–90% of the voltage appearing at the electrode under fault conditions is dropped across the earth in the first 2.5 to 3 m. This is particularly dangerous where livestock is present, as the hind and fore legs of an animal can be respectively inside and outside the resistance area: a PD of 25 V can be lethal! One method of overcoming this problem is to house the electrode in a pit below ground level (Figure 7) as this prevents voltages appearing at ground level.

Earthing in the IEE Regulations

(IEE Regs Chapter 4)
(IEE Regs 413–02–01 to 28)

In the preceding pages we have briefly discussed the reasons for, and the importance and methods of, earthing. Let us now examine the subject in relation to the IEE Regulations.

Contact with metalwork made live by a fault is called *indirect contact*. One popular method of providing some measure of protection against such contact is by earthed equipotential bonding and automatic disconnection of supply (E.E.B.A.D.S.). This entails the bonding together and connection to earth of:

1 All metalwork associated with electrical apparatus and systems, termed exposed conductive parts. Examples include conduit, trunking and the metal cases of apparatus.
2 All metalwork liable to introduce a potential including earth potential, termed extraneous conductive parts. Examples are gas, oil and water pipes, structural steelwork, radiators, sinks and baths.

The conductors used in such connections are called *protective conductors*, and they can be further subdivided into:

1 Circuit protective conductors, for connecting exposed conductive parts to the main earthing terminal.
2 Main equipotential bonding conductors, for bonding together main incoming services, structural steelwork etc.
3 Other equipotential bonding conductors, for bonding together sinks, baths, taps etc.
4 Supplementary bonding conductors for bonding exposed conductive parts and extraneous conductive parts, when circuit disconnection times cannot be met, or, in special locations, such as bathrooms, swimming pools etc.

The effect of all this bonding is to create a zone in which all metalwork of different services and systems will, even under fault conditions, be at a substantially equal potential. If, added to this, there is a low-resistance earth return path, the protection should operate fast enough to prevent danger. [IEE Reg 413–02–04.]

The resistance of such an earth return path will depend upon the system (see the next section), either TT, TN–S or TN–C–S (IT systems will not be discussed here, as they are extremely rare and unlikely to be encountered by the average contractor).

Earthing systems

(IEE Regs Definitions (Systems))

These have been designated in the IEE Regulations using the letters T, N, C and S. These letters stand for:

T terre (French for earth) and meaning a direct connection to earth
N neutral
C combined
S separate

When these letters are grouped they form the classification of a type of system. The first letter in such a classification denotes how the supply source is earthed. The second denotes how the metalwork of an installation is earthed. The third and fourth indicate the functions of neutral and protective conductors. Hence:

1 A TT system has a direct connection of the supply source neutral to earth and a direct connection of the installation metalwork to earth. An example is an overhead line supply with earth electrodes, and the mass of earth as a return path (Figure 8).

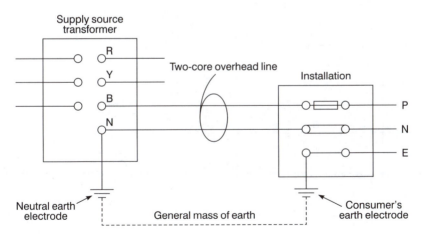

Figure 8 TT system

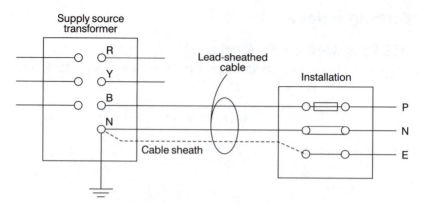

Figure 9   TN–S system

2 A TN–S system has the supply source neutral directly connected
   to earth, the installation metalwork connected to the earthed
   neutral of the supply source via the lead sheath of the supply
   cable and the neutral and protective conductors throughout the
   whole system performing separate functions (Figure 9).
3 A TN–C–S system is as the TN–S but the supply cable sheath is
   also the neutral, i.e. it forms a combined earth/neutral conductor
   known as a PEN (protective earthed neutral) conductor (Figure 10).

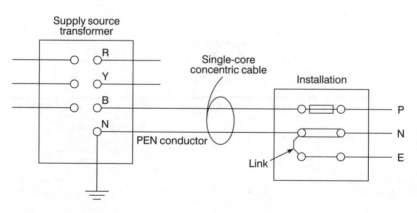

Figure 10   TN–C–S system

The installation earth and neutral are separate conductors. This system is also known as PME (protective multiple earthing).

Note that only single-phase systems have been shown, for simplicity.

Summary

In order to reduce the risk of serious electric shock, it is important to provide a path for earth leakage currents to operate the circuit protection, and to endeavour to maintain all metalwork at a substantially equal potential. This is achieved by bonding together metalwork of electrical and non-electrical systems to earth. The path for leakage currents would then be via the earth itself in TT systems or by a metallic return path in TN–S or TN–C–S systems.

Earth fault loop impedance

(IEE Regs Definitions)

As we have seen, circuit protection should operate in the event of a direct fault from phase to earth. The speed of operation of the protection is of extreme importance and will depend on the magnitude of the fault current, which in turn will depend on the impedance of the earth fault loop path, Z_s.

Figure 11 shows this path. Starting at the fault, the path comprises:

1 The circuit protective conductor (CPC).
2 The consumer's earthing terminal and earth conductor.
3 The return path, either metallic or earth.
4 The earthed neutral of the supply transformer.
5 The transformer winding.
6 The phase conductor from the transformer to the fault.

Figure 12 is a simplified version of this path. We have:

$$Z_s = Z_e + R_1 + R_2$$

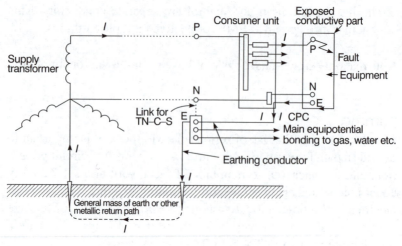

Figure 11

where Z_s is the actual total loop impedance, Z_e is the loop impedance external to the installation, R_1 is the resistance of the phase
conductor, and R_2 is the resistance of the CPC. We also have:

$$I = U_{0c}/Z_s$$

where I is the fault current and U_{0c} is the voltage at the supply
transformer (assumed to be 240 V) (Appendix 3 BS 7671).

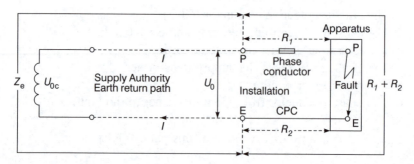

Figure 12

Determining the value of total loop impedance

(IEE Regs 413-02-04)

The IEE Regulations require that when the general characteristics of an installation are assessed, the loop impedance Z_e external to the installation shall be ascertained.

This may be measured in existing installations using a phase-to-earth loop impedance tester. However, when a building is only at the drawing board stage it is clearly impossible to make such a measurement. In this case, we have three methods available to assess the value of Z_e:

1 Determine it from details (if available) of the supply transformer, the main distribution cable and the proposed service cable; or
2 Measure it from the supply intake position of an adjacent building having service cable of similar size and length to that proposed; or
3 Use maximum likely values issued by the supply authority as follows:

> TT system: 21 Ω maximum
> TN-S system: 0.80 Ω maximum
> TN-C-S system: 0.35 Ω maximum.

Method 1 will be difficult for anyone except engineers. Method 3 can, in some cases, result in pessimistically large cable sizes. Method 2, if it is possible to be used, will give a closer and more realistic estimation of Z_e. However, if in any doubt, use method 3.

Having established a value for Z_e, it is now necessary to determine the impedance of that part of the loop path internal to the installation. This is, as we have seen, the resistance of the phase conductor plus the resistance of the CPC, i.e. $R_1 + R_2$. Resistances of copper conductors may be found from manufacturers' information which gives values of resistance/metre for copper and aluminium conductors at 20 °C in mΩ/m. The following table gives resistance values for copper conductors up to 35 mm^2.

A 25 mm^2 phase conductor with a 4 mm^2 CPC has $R_1 = 0.727$ and $R_2 = 4.61$, giving $R_1 + R_2 = 0.727 + 4.61 = 5.337$ mΩ/m. So, having established a value for $R_1 + R_2$, we must now multiply it by the length of the run and divide by 1000 (the values given are in mΩ/m). However, this final value is based on a temperature of 20 °C, but when the conductor is fully loaded its temperature will increase. In order to determine the value of resistance at conductor operating temperature, a multiplier is used.

This multiplier, applied to the 20 °C value of resistance is determined from the following formula:

$$R_t = R_{20}\{1 + \alpha_{20}(\theta° - 20°)\}$$

Where R_t = the resistance at conductor operating temperature
R_{20} = the resistance at 20 °C
α_{20} = the 20 °C temperature coefficient of copper, 0.004 Ω/Ω/°C
$\theta°$ = the conductor operating temperature.

Clearly, the multiplier is $\{1 + \alpha_{20}(\theta° - 20°)\}$.

So, for a 70 °C, polyvinyl chloride (PVC) insulated conductor (Table 54C IEE Regulations), the multiplier becomes:

$$\{1 + 0.004(70° - 20°)\} = 1.2$$

And for a 90 °C XLPE type cable it becomes:

$$\{1 + 0.004(90° - 20°)\} = 1.28$$

Hence, for a 20 m length of 70 °C PVC insulated 16 mm² phase conductor with a 4 mm² CPC, the value of $R_1 + R_2$ would be:

$$R_1 + R_2 = (1.15 + 4.61) \times 20 \times 1.2/1000 = 0.138 \ \Omega$$

We are now in a position to determine the total earth fault loop impedance Z_s from:

$$Z_s = Z_e + R_1 + R_2$$

Table 1   Resistance of copper conductors in at 20 °C

Conductor CSA (mm²)	Resistance (mΩ/m)
1.0	18.1
1.5	12.1
2.5	7.41
4.0	4.61
6.0	3.08
10.0	1.83
16.0	1.15
25.0	0.727
35.0	0.524

As previously mentioned, this value of Z_s should be as low as possible to allow enough fault current to flow to operate the protection as quickly as possible. Tables 41B1, B2 and D of the IEE Regulations give maximum values of loop impedance for different sizes and types of protection for both socket outlet circuits, and circuits feeding fixed equipment.

Provided that the actual values calculated do not exceed those tabulated, socket outlet circuits will disconnect under earth fault conditions in 0.4 s or less, and circuits feeding fixed equipment in 5 s or less. The reasoning behind these different times is based on the time that a faulty circuit can reasonably be left uninterrupted. Hence, socket outlet circuits from which hand-held appliances may be used, clearly present a greater shock risk than circuits feeding fixed equipment. BS 7671 does not mention dis-connection times in bathrooms, but it would be wise to assume the lower disconnection time of 0.4 s, as bathrooms etc. are hazardous areas.

It should be noted that these times, i.e. 0.4 s and 5 s, do not indicate the duration that a person can be in contact with a fault. They are based on the probable chances of someone being in contact with exposed or extraneous conductive parts at the precise moment that a fault develops.

See also Tables 41A, 604A and 605A of the IEE Regulations.

Example 1

Let us now have a look at a typical example of, say, a shower circuit run in an 18 m length of $6.0\,\text{mm}^2$ (6242Y) twin cable with CPC, and protected by a 30 A BS 3036 semi- enclosed rewirable fuse. A $6.0\,\text{mm}^2$ twin cable has a $2.5\,\text{mm}^2$ CPC. We will also assume that the external loop impedance Z_e, is measured as $0.27\,\Omega$. Will there be a shock risk if a phase-to-earth fault occurs?

The total loop impedance $Z_s = Z_e + R_1 + R_2$. We are given $Z_e = 0.27\,\Omega$. For a $6.0\,\text{mm}^2$ phase conductor with a $2.5\,\text{mm}^2$ CPC, $R_1 + R_2$ is $10.49\,\text{m}\Omega/\text{m}$. Hence, with a multiplier of 1.2 for 70 °C PVC,

$$\text{total } R_1 + R_2 = 18 \times 10.49 \times 1.2/1000 = 0.23\,\Omega$$

Therefore, $Z_s = 0.27 + 0.23 = 0.53\,\Omega$. This is less than the $1.14\,\Omega$ maximum given in Table 41B1 for a 30 A BS 3036 fuse. Hence, the protection will disconnect the circuit in less than 0.4 s. In fact it will disconnect in less than 0.1 s, the determination of this time will be dealt with in Chapter 5.

Example 2

Consider now, a more complex installation, and note how the procedure remains unchanged.

In this example, a 3-phase motor is fed using $25\,\text{mm}^2$ single PVC conductors in trunking, the CPC being $2.5\,\text{mm}^2$. The circuit is protected by BS 1361 45 A fuses in a distribution fuseboard. The distribution circuit or sub-main feeding this fuseboard comprises $70\,\text{mm}^2$ PVC singles in trunking with a $25\,\text{mm}^2$ CPC the protection being by BS 88 160 A fuses. The external loop impedance Z_e has a measured value of $0.2\,\Omega$. Will this circuit arrangement comply with the shock risk constraints?

The formula $Z_s = Z_e + R_1 + R_2$ must be extended, as the $(R_1 + R_2)$ component comprises both sub-main and motor circuit, it therefore becomes:

$$Z_s = Z_e + (R_1 + R_2)_1 + (R_1 + R_2)_2$$

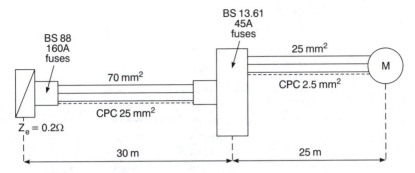

Figure 13

Distribution circuit $(R_1+R_2)_1$

This comprises 30 m of 70 mm² single-phase conductor and 30 m of 25 mm² CPC. Typical values for conductors over 35 mm² are shown in Table 2.

As an alternative we can use our knowledge of the relationship between conductor resistance and area, e.g. a 10 mm² conductor has approximately 10 times less resistance than a 1 mm² conductor:

$$10\,\text{mm}^2 \text{ resistance } = 1.83\,\text{m}\Omega/\text{m}$$
$$1.0\,\text{mm}^2 \text{ resistance } = 18.1\,\text{m}\Omega/\text{m}$$

Hence a 70 mm² conductor will have a resistance approximately half that of a 35 mm² conductor.

$$35\,\text{mm}^2 \text{ resistance } = 0.524\,\text{m}\Omega/\text{m}$$
$$\therefore 70\text{ mm}^2 \text{ resistance } = \frac{0.524}{2} = 0.262\,\text{m}\Omega/\text{m}$$

which compares well with the value given in Table 2.

$$25\,\text{mm}^2 \text{ CPC resistance } = 0.727\,\text{m}\Omega/\text{m}$$

so the distribution circuit

$$(R_1+R_2)_1 = 30 \times (0.262+0.727) \times 1.2/1000 = 0.035\,\Omega.$$

Table 2

Area of conductor (mm²)	Resistance (mΩ/m)	
	Copper	Aluminium
50	0.387	0.641
70	0.263	0.443
95	0.193	0.320
120	0.153	0.253
185	0.0991	0.164
240	0.0754	0.125
300	0.0601	0.1

Hence $Z_s = Z_e + (R_1 + R_2)_1 = 0.2 + 0.035 = 0.235\,\Omega$ which is less than the Z_s, maximum of $0.267\,\Omega$ quoted for a 160 A BS 88 fuse in Table 41D of the Regulations.

Motor circuit $(R_1+R_2)_2$
Here we have 25 m of 25 mm² conductor with 25 m of 2.5 mm² CPC. Hence:

$$(R_1 + R_2)_2 = 25 \times (0.727 + 7.41) \times 1.2/1000$$

$$= 0.24\,\Omega$$

$$\therefore \text{Total } Z_s = Z_e + (R_1 + R_2)_1 + (R_1 + R_2)_2$$

$$= 0.2 + 0.035 + 0.24$$

$$= 0.48\,\Omega$$

which is less than the Z_s maximum of $1.0\,\Omega$ quoted for a 45 A BS 1361 fuse from Table 41D of the Regulations.

Hence we have achieved compliance with the shock-risk constraints.

Residual current devices

(IEE Regs 531, 412-06, 413-02-15 and 471-16-01)

We have seen how very important is the total earth loop impedance Z_s in the reduction of shock risk. However, in TT systems where the mass of earth is part of the fault path, the maximum values of Z_s given in Tables 41 B1, B2 and D of the Regulations may be hard to satisfy. Added to this, climatic conditions will alter the resistance of the earth in such a way that Z_s may be satisfactory in wet weather but not in very dry.

The IEE Regulations recommend therefore that protection for circuits in a TT system be achieved by residual current devices (RCD's), such that the product of the residual operating current and the loop impedance will not exceed a figure of 50 V. Residual current breakers (RCBs), residual current circuit breakers (RCCBs) and RCDs are one and the same thing. RCBO's are combined circuit breakers (CB's) and RCD's in one unit.

For construction sites and agricultural environments this value is reduced to 25 V.

Principle of operation of an RCD

Figure 14 illustrates the construction of an RCD. In a healthy circuit the same current passes through the phase coil, the load, and back through the neutral coil. Hence the magnetic effects of phase and neutral currents cancel out.

In a faulty circuit, either phase to earth or neutral to earth, these currents are no longer equal. Therefore the out-of-balance current produces some residual magnetism in the core. As this magnetism is alternating, it links with the turns of the search coil, inducing an EMF in it. This EMF in turn drives a current through the trip coil, causing operation of the tripping mechanism.

It should be noted that a phase-to-neutral fault will appear as a load, and hence the RCD will not operate for this fault.

A three-phase RCD works on the same out of balance principle, in this case the currents flowing in the three phases when they are all equal, sum to zero, hence there is no resultant magnetism. Even

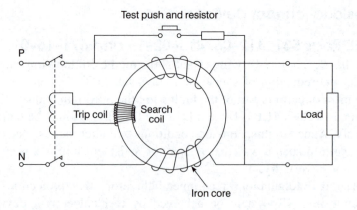

Figure 14a   Residual current device

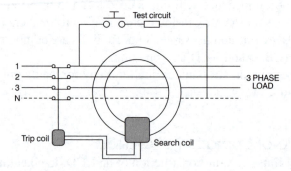

Figure 14b   Three-phase RCD

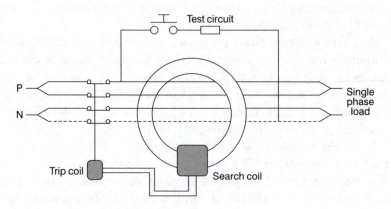

Figure 14c   Connections for single phase

if they are unequal, the out of balance current flows in the neutral which cancels out this out of balance current. Figure 14(b) shows the arrangement of a three-phase RCD, and Figure 14(c), how it can be connected for use on single-phase circuits.

Nuisance tripping

Certain appliances such as cookers, water heaters and freezers tend to have, by the nature of their construction and use, some leakage currents to earth. These are quite normal, but could cause the operation of an RCD protecting an entire installation. This can be overcome by using split-load consumer units, where socket outlet circuits are protected by a 30 mA RCD, leaving all other circuits controlled by a normal mains switch. Better still, especially in TT systems, is the use of a 100 mA RCD for protecting circuits other than socket outlets.

Modern developments in MCB, RCD and consumer unit design now make it easy to protect any individual circuit with a combined MCB/RCD (RCBO), making the use of split-load boards unnecessary.

One area where the use of 30 mA RCDs is required is in the protection of socket outlets intended for the connection of portable appliances for use outside the main equipotential zone. Hence, socket outlets in garages or even within the main premises which are likely to be used for supplying portable tools such as lawn mowers and hedge trimmers must be protected by an RCD rated at 30 mA or less. All other equipment outside the main equipotential zone should, in the event of an earth fault, disconnect in 0.4 s.

An exception to the RCD requirement is where fixed equipment is connected to the supply via a socket outlet, provided that some means of preventing the socket outlet being used for hand-held appliances is ensured.

Supplementary bonding

(IEE Regs 413-02-15(i), 413-02-27 and 28 and Section 547-03)

This is perhaps the most debated topic in the IEE Regulations. The confusion may have arisen because of a lack of understanding of earthing and bonding. Hopefully, this chapter will rectify the situation. In general the only Supplementary bonding required in a domestic premises is in a bathroom.

By now we should know why bonding is necessary; the next question, however, is to what extent bonding should be carried out. This is perhaps answered best by means of question and answer examples:

1 Q: Why do I need to bond the hot and cold taps and a metal kitchen sink together? Surely they are all joined anyway?

A: In most sinks the holes for connection of the taps are usually surrounded by a plastic insert which tends to insulate the taps from the sink. The sink is a large metallic part which is in the zone & therefore not an extraneous conductive part. The hot and cold taps are both parts of different systems and could originate from outside the equipotential zone. These, therefore, could be extraneous conductive parts and may need to be bonded together, although there is no specific requirement in BS 7671 to do this.

2 Q: Do I have to bond radiators in a premises to, say, metal clad switches or socket outlets etc.?

A: Supplementary bonding is only necessary when extraneous conductive parts are simultaneously accessible with exposed conductive parts and the disconnection time for the circuit concerned cannot be achieved. In these circumstances the bonding conductor should have a resistance $R \leq 50/I_a$, where I_a is the operating current of the protection.

3 Q: Do I need to bond metal window frames?

A: In general, no. Apart from the fact that most window frames will not introduce a potential from anywhere, the part of the window most likely to be touched is the opening portion, to which it would not be practicable to bond. There may be a case for the bonding of patio doors, which could be considered earthy with rain running from the lower portion to the earth. However, once again the part most likely to be touched is the sliding section, to which it is not possible to bond. In any case there would need to be another simultaneously accessible part to warrant considering any bonding.

4 Q: What about bonding in bathrooms?

A: Bathrooms are particularly hazardous areas with regard to shock risk, as body resistance is drastically reduced when wet. Hence, supplementary bonding between exposed conductive parts must be carried out in addition to their existing CPCs. Also of course, taps and metal baths may need bonding together, and to other extraneous and exposed conductive parts. It may be of interest to note that in older premises a toilet basin may be connected into a cast iron collar which then tees outside into a cast iron soil pipe. This arrangement will clearly introduce earth potential into the bathroom, and hence the collar should be bonded to any simultaneously accessible conductive parts. This may require an unsightly copper earth strap.

5 Q: What size of bonding conductors should I use?

A: Main equipotential bonding conductors should be not less than half the size of the main earthing conductor, subject to a minimum of $6.0 \, \text{mm}^2$ or, where PME (TNCS) conditions are present, $10.0 \, \text{mm}^2$. For example, most new domestic installations now have a $16.0 \, \text{mm}^2$ earthing conductor, so all main bonding will be in $10.0 \, \text{mm}^2$. Supplementary bonding conductors are subject to a minimum of $2.5 \, \text{mm}^2$ if mechanically protected or $4.0 \, \text{mm}^2$ if not. However, if these bonding conductors are connected to exposed conductive parts, they must be the same size as the CPC connected to the exposed conductive part, once again subject to the minimum sizes mentioned. It is sometimes difficult to protect a bonding conductor mechanically throughout its length, and especially at terminations, so it is perhaps better to use $4.0 \, \text{mm}^2$ as the minimum size.

6 Q: Do I have to bond free-standing metal cabinets, screens, workbenches etc.?

A: No. These items will not introduce a potential into the equipotential zone from outside, and cannot therefore be regarded as extraneous conductive parts.

7 Q: What are the bonding requirements for plumbing installations that incorporate plastic pipes.

A: There is an increasing amount of plastic plumbing installations being used in modern houses for both domestic hot and cold water and C.H. systems. If the pipework is plastic but terminates in copper at taps, radiators etc. NO bonding is needed.

The Faraday cage

In one of his many experiments, Michael Faraday (1791–1867) placed an assistant in an open-sided cube which was then covered in a conducting material and insulated from the floor. When this cage arrangement was charged to a high voltage, the assistant found that he could move freely within it touching any of the sides, with no adverse effects. Faraday had, in fact, created an equipotential zone, and of course in a correctly bonded installation, we live and/or work in Faraday cages!

Problems

1 What is the resistance of a 10 m length of 6.0 mm^2 copper phase conductor if the associated CPC is 1.5 mm^2?

2 What is the length of a 6.0 mm^2 copper phase conductor with a 2.5 mm^2 CPC if the overall resistance is 0.189 Ω?

3 If the total loop impedance of a circuit under operating conditions is 0.96 Ω and the cable is a 20 m length of 4.0 mm^2 copper with a 1.5 mm^2 CPC, what is the external loop impedance?

4 Will there be a shock risk if a double socket outlet, fed by a 23 m length of 2.5 mm^2 copper conductor with a 1.5 mm^2 CPC, is protected by a 20 A BS 3036 rewirable fuse and the external loop impedance is measured as 0.5 Ω?

5 A cooker control unit incorporating a socket outlet is protected by a 32 A BS 88 fuse, and wired in 6.0 mm^2 copper with a 2.5 mm^2 CPC. The run is some 25 m and the external loop impedance of the TN–S system is not known. Is there a shock risk, and if so, how could it be rectified?

3
PROTECTION

(Relevant IEE parts, chapters and sections
Part 4; Chapters 41, 42, 43, 44, 45, 48; Section 471,
473, 514, 522, 543, 547)

Definitions used in this chapter

Arm's reach A zone of accessibility to touch, extending from any point on a surface where persons usually stand or move about, to the limits which a person can reach with his hand in any direction without assistance.

Barrier A part providing a defined degree of protection against contact with live parts, from any usual direction.

Class 2 equipment Equipment in which protection against electric shock does not rely on basic insulation only, but in which additional safety precautions such as supplementary insulation are provided. There is no provision for the connection of exposed metalwork of the equipment to a protective conductor, and no reliance upon precautions to be taken in the fixed wiring of the installation.

Circuit protective conductor A protective conductor connecting exposed conductive parts of equipment to the main earthing terminal.

Protection

Design current   The magnitude of the current intended to be carried by a circuit in normal service.

Direct contact   Contact of persons or livestock with live parts.

Enclosure   A part providing an appropriate degree of protection of equipment against certain external influences and a defined degree of protection against contact with live parts from any direction.

Exposed conductive part   A conductive part of equipment which can be touched and which is not a live part but which may become live under fault conditions.

External influence   Any influence external to an electrical installation which affects the design and safe operation of that installation.

Extraneous conductive part   A conductive part liable to introduce a potential, generally earth potential, and not forming part of the electrical installation.

Fault current   A current resulting from a fault.

Fixed equipment   Equipment fastened to a support or otherwise secured in a specific location.

Indirect contact   Contact of persons or livestock with exposed conductive parts which have become live under fault conditions.

Isolation   Cutting off an electrical installation, a circuit or an item of equipment from every source of electrical energy.

Insulation   Suitable non-conductive material enclosing, surrounding or supporting a conductor.

Live part   A conductor or conductive part intended to be energized in normal use, including a neutral conductor but, by convention, not a PEN conductor.

Obstacle   A part preventing unintentional contact with live parts but not preventing deliberate contact.

Overcurrent   A current exceeding the rated value. For conductors the rated value is the current-carrying capacity.

Overload An overcurrent occurring in a circuit which is electrically sound.

Residual current device An electromechanical switching device or association of devices intended to cause the opening of the contacts when the residual current attains a given value under specified conditions.

Short-circuit current An overcurrent resulting from a fault of negligible impedance between live conductors having a difference of potential under normal operating conditions.

Skilled person A person with technical knowledge or sufficient experience to enable him to avoid the dangers which electricity may create.

What is protection?

The meaning of the word 'protection', as used in the electrical industry, is no different to that in everyday use. People protect themselves against personal or financial loss by means of insurance and from injury or discomfort by the use of the correct protective clothing. They further protect their property by the installation of security measures such as locks and/or alarm systems. In the same way, electrical systems need:

1 To be protected against mechanical damage, the effects of the environment and electrical overcurrents; and
2 To be installed in such a fashion that persons and/or livestock are protected from the dangers that such an electrical installation may create.

Let us now look at these protective measures in more detail.

Protection against mechanical damage

The word 'mechanical' is somewhat misleading in that most of us associate it with machinery of some sort. In fact a serious electrical overcurrent left uninterrupted for too long can cause distortion of conductors and degradation of insulation; both of these effects are considered to be mechanical damage.

However, let us start by considering the ways of preventing mechanical damage by physical impact and the like.

Cable construction

A cable comprises one or more conductors each covered with an insulating material. This insulation provides protection from shock by direct contact and prevents the passage of leakage currents between conductors.

Clearly, insulation is very important and in itself should be protected from damage. This may be achieved by covering the insulated conductors with a protective sheathing during manufacture, or by enclosing them in conduit or trunking at the installation stage.

The type of sheathing chosen and/or the installation method will depend on the environment in which the cable is to be installed. For example, metal conduit with PVC singles or mineral-insulated (MI) cable would be used in preference to PVC-sheathed cable clipped direct, in an industrial environment. Figure 15 shows the effect of physical impact on MI cable.

Protection against corrosion

Mechanical damage to cable sheaths and metalwork of wiring systems can occur through corrosion, and hence care must be taken to choose corrosion-resistant materials and to avoid contact between dissimilar metals in damp situations.

Protection against thermal effects

This is the subject of Chapter 42 of the IEE Regulations. Basically, it requires common-sense decisions regarding the placing of fixed equipment, such that surrounding materials are not at risk from damage by heat.

Added to these requirements is the need to protect persons from burns by guarding parts of equipment liable to exceed temperatures listed in Table 42A of the Regulations.

Figure 15 Mineral-insulated cable. On impact, all parts including the conductors are flattened, and a proportionate thickness of insulation remains between conductors, and conductors and sheath, without impairing the performance of the cable at normal working voltages

Polyvinyl chloride

PVC is a thermoplastic polymer widely used in electrical installation work for cable insulation, conduit and trunking. General-purpose PVC is manufactured to the British Standard BS 6746.

PVC in its raw state is a white powder; it is only after the addition of plasticizers and stabilizers that it acquires the form that we are familiar with.

Degradation

All PVC polymers are degraded or reduced in quality by heat and light. Special stabilizers added during manufacture help to retard this degradation at high temperatures. However, it is recommended that PVC-sheathed cables or thermoplastic fittings for luminaries (light fittings) should not be installed where the temperature is likely to rise above 60 °C. Cables insulated with high-temperature PVC (up to 80 °C) should be used for drops to lampholders and entries into batten-holders. PVC conduit and trunking should not be used in temperatures above 60 °C.

Embrittlement and cracking

PVC exposed to low temperatures becomes brittle and will easily crack if stressed. Although both rigid and flexible, PVC used in cables and conduit can reach as low as 5°C without becoming brittle, it is recommended that general-purpose PVC insulated cables should not be installed in areas where the temperature is likely to be consistently below 0°C, and that PVC-insulated cable should not be handled unless the ambient temperature is above 0°C and unless the cable temperature has been above 0°C for at least 24 hours.

Where rigid PVC conduit is to be installed in areas where the ambient temperature is below −5°C but not lower than −25°C, type B conduit manufactured to BS 4607 should be used.

When PVC-insulated cables are installed in loft spaces insulated with polystyrene granules, contact between the two polymers can cause the plasticizer in the PVC to migrate to the granules. This causes the PVC to harden and although there is no change in the electrical properties, the insulation may crack if disturbed.

External influences

Appendix 5 of the IEE Regulations classifies external influences which may affect an installation. This classification is divided into three sections, the environment (A), how that environment is utilized (B) and construction of buildings (C). The nature of any influence within each section is also represented by a letter, and the level of influence by a number. The following gives examples of the classification:

Environment	Utilization	Building
Water	Capability	Materials
AD6 Waves	**BA3** Handicapped	**CA1** Non-combustible

With external influences included on drawings and in specifications, installations and materials used can be designed accordingly.

Protection against ingress of solid objects and liquid

In order to protect equipment from damage by foreign bodies or liquid, and also to prevent persons from coming into contact with live or moving parts, such equipment is housed inside an enclosure.

The degree of protection offered by such an enclosure is the subject of BS EN 60529, commonly known as the IP code, part of which is as shown in the accompanying table. It will be seen from this table that, for instance, an enclosure to IP56 is dustproof and waterproof.

IP codes

First numeral: mechanical protection

0 No protection of persons against contact with live or moving parts inside the enclosure. No protection of equipment against ingress of solid foreign bodies.

1 Protection against accidental or inadvertent contact with live or moving parts inside the enclosure by a large surface of the human body, for example, a hand, but not protection against deliberate access to such parts. Protection against ingress of large solid foreign bodies.

2 Protection against contact with live or moving parts inside the enclosure by fingers. Protection against ingress of medium-size solid foreign bodies.

3 Protection against contact with live or moving parts inside the enclosures by tools, wires or such objects of thickness greater than 2.5 mm. Protection against ingress of small foreign bodies.

4 Protection against contact with live or moving parts inside the enclosure by tools, wires or such objects of thickness greater than 1 mm. Protection against ingress of small solid foreign bodies.

5 Complete protection against contact with live or moving parts inside the enclosure. Protection against harmful deposits of dust. The ingress of dust is not totally prevented, but dust cannot enter in an amount sufficient to interfere with satisfactory operation of the equipment enclosed.

6 Complete protection against contact with live or moving parts inside the enclosures. Protection against ingress of dust.

Protection

Second numeral: liquid protection

0 No protection.

1 Protection against drops of condensed water. Drops of condensed water falling on the enclosure shall have no harmful effect.

2 Protection against drops of liquid. Drops of falling liquid shall have no harmful effect when the enclosure is tilted at any angle up to 15° from the vertical.

3 Protection against rain. Water falling in rain at an angle equal to or smaller than 60° with respect to the vertical shall have no harmful effect.

4 Protection against splashing. Liquid splashed from any direction shall have no harmful effect.

5 Protection against water jets. Water projected by a nozzle from any direction under stated conditions shall have no harmful effect.

6 Protection against conditions on ships' decks (deck with watertight equipment). Water from heavy seas shall not enter the enclosures under prescribed conditions.

7 Protection against immersion in water. It must not be possible for water to enter the enclosure under stated conditions of pressure and time.

8 Protection against indefinite immersion in water under specified pressure. It must not be possible for water to enter the enclosure.

X Indicates no *specified* protection.

The most commonly quoted IP codes in the Regulations are IP2X, IP4X and IPXXB. (The X denotes that no protection is specified, *not* that no protection exists.)

Hence, IP2X means that an enclosure can withstand the ingress of medium-sized solid foreign bodies (12.5 mm diameter), and a jointed test finger, known affectionately as the British Standard finger! IPXXB denotes protection against the test finger only, and IP4X indicates protection against small foreign bodies and a 1 mm diameter test wire.

IEE Regulations 522–01 to 12 give details of the types of equipment, cables, enclosure etc. that may be selected for certain environmental conditions, e.g an enclosure housing equipment in an AD8 environment (under water) would need to be to IPX8.

Protection against electric shock

(IEE Regs Chapter 41 and Section 471)

There are two ways of receiving an electric shock: by direct contact, and by indirect contact. It is obvious that we need to provide protection against both of these conditions.

Protection against direct contact

(IEE Regs Section 412 and Regs 471-04 to 07)

Clearly, it is not satisfactory to have live parts accessible to touch by persons or livestock. The IEE Regulations recommend five ways of minimizing this danger:

1 By covering the live part or parts with insulation which can only be removed by destruction, e.g. cable insulation.
2 By placing the live part or parts behind a barrier or inside an enclosure providing protection to at least 1P2X or IPXXB. In most cases, during the life of an installation it becomes necessary to open an enclosure or remove a barrier. Under these circumstances, this action should only be possible by the use of a key or tool, e.g. by using a screwdriver to open a junction box. Alternatively, access should only be gained after the supply to the live parts has been disconnected, e.g. by isolation on the front of a control panel where the cover cannot be removed until the isolator is in the 'off' position. An intermediate barrier of at least IP2X or IPXXB will give protection when an enclosure is opened: a good example of this is the barrier inside distribution fuseboards, preventing accidental contact with incoming live feeds.
3 By placing obstacles to prevent unintentional approach to or contact with live parts. This method must only be used where skilled persons are working.
4 By placing out of arm's reach: for example, the high level of the bare conductors of travelling cranes.
5 By using an RCD. Whilst not permitted as the sole means of protection, this is considered to reduce the risk associated with direct contact, provided that one of the other methods just

mentioned is applied, and that the RCD has a rated operating current, $I_{\Delta n}$ of not more than 30 mA and an operating time not exceeding 40 ms at 5 times $I_{\Delta n}$, i.e. 150 mA.

Protection against indirect contact

(IEE Regs Section 413 and Regs 471–08 to 12)

The IEE Regulations suggest five ways of protecting against indirect contact. One of these, earthed equipotential bonding and automatic disconnection of supply, has already been discussed in Chapter 2. The other methods are as follows.

Use of class 2 equipment

Often referred to as double-insulated equipment, this is typical of modern appliances where there is no provision for the connection of a CPC. This does not mean that there should be no exposed conductive parts and that the casing of equipment should be of an insulating material; it simply indicates that live parts are so well insulated that faults from live to conductive parts cannot occur.

Non-conducting location

This is basically an area in which the floor, walls and ceiling are all insulated. Within such an area there must be no protective conductors, and socket outlets will have no earthing connections.

It must not be possible simultaneously to touch two exposed conductive parts, or an exposed conductive part and an extraneous conductive part. This requirement clearly prevents shock current passing through a person in the event of an earth fault, and the insulated construction prevents shock current passing to earth.

Earth-free local equipotential bonding

This is, in essence, a Faraday cage, where all metal is bonded together but *not* to earth. Obviously great care must be taken when entering such a zone in order to avoid differences in potential between inside and outside.

The areas mentioned in this and the previous method are very uncommon. Where they do exist, they should be under constant supervision to ensure that no additions or alterations can lessen the protection intended.

Electrical separation

This method relies on a supply from a safety source such as an isolating transformer to BS EN 60742 which has no earth connection on the secondary side. In the event of a circuit that is supplied from a source developing a live fault to an exposed conductive part, there would be no path for shock current to flow: see Figure 16.

Once again, great care must be taken to maintain the integrity of this type of system, as an inadvertent connection to earth, or interconnection with other circuits, would render the protection useless.

Exemptions

(IEE Regs 471-13)

As with most sets of rules and regulations, there are certain areas which are exempt from the requirements. These are listed quite

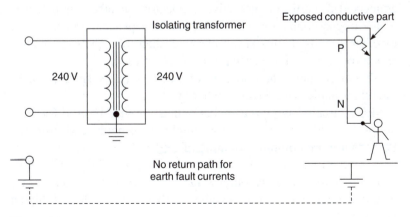

Figure 16

clearly in IEE Regulations 471–13, and there is no point in repeating them all here. However, one example is the dispensing of the need to earth exposed conductive parts such as small fixings, screws and rivets, provided that they cannot be touched or gripped by a major part of the human body (not less than 50 mm by 50 mm), and that it is difficult to make and maintain an earth connection.

Protection against direct and indirect contact

(IEE Regs Section 411)

So far we have dealt separately with direct and indirect contact. However, we can protect against both of these conditions with the following methods.

Separated extra low voltage (SELV)

This is simply extra low voltage (less than 50 V AC) derived from a safety source such as a class 2 safety isolating transformer to BS 3535; or a motor generator which has the same degree of isolation as the transformer; or a battery or diesel generator; or an electronic device such as a signal generator.

Live or exposed conductive parts of SELV circuits should not be connected to earth, or protective conductors of other circuits, and SELV circuit conductors should ideally be kept separate from those of other circuits. If this is not possible, then the SELV conductors should be insulated to the highest voltage present.

Obviously, plugs and sockets of SELV circuits should not be interchangeable with those of other circuits.

SELV circuits supplying socket outlets are mainly used for hand lamps or soldering irons, for example, in schools and colleges. Perhaps a more common example of a SELV circuit is a domestic bell installation, where the transformer is to BS EN 60742. Note that bell wire is usually only suitable for 50–60 V, which means that it should not be run together with circuit cables of higher voltages.

Reduced low voltage systems

(IEE Regs 471-15)

The Health and Safety Executive accepts that a voltage of 65 V to earth, three-phase, or 55 V to earth, single-phase, will give protection against severe electric shock. They therefore recommend that portable tools used on construction sites etc. be fed from a 110 V centre-tapped transformer to BS 4343. Figure 17 shows how 55 V is derived. Earth fault loop impedance values for these systems may be taken from Table 471A of the Regulations.

Protection against overcurrent

(IEE Regs Chapter 43 and Definitions)

An overcurrent is a current greater than the rated current of a circuit. It may occur in two ways:

1 As an overload current; or
2 As a short-circuit or fault current.

These conditions need to be protected against in order to avoid damage to circuit conductors and equipment. In practice, fuses and circuit breakers will fulfil both of these needs.

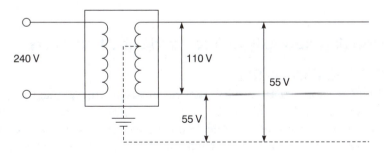

Figure 17

Overloads

Overloads are overcurrents occurring in healthy circuits. They may be caused, for example, by faulty appliances or by surges due to motors starting or by plugging in too many appliances in a socket outlet circuit.

Short circuits

A short-circuit current is the current that will flow when a 'dead short' occurs between live conductors (phase to neutral for single phase; phase to phase for three–phase). Prospective short-circuit current is the same, but the term is usually used to signify the value of short-circuit current at fuse or circuit breaker positions.

Prospective short-circuit current is of great importance. However, before discussing it or any other overcurrent further, it is perhaps wise to refresh our memories with regard to fuses and circuit breakers and their characteristics.

Fuses and circuit breakers

As we all know, a fuse is the weak link in a circuit which will break when too much current flows, thus protecting the circuit conductors from damage.

There are many different types and sizes of fuse, all designed to perform a certain function. The IEE Regulations refer to only four of these: BS 3036, BS 88, BS 1361 and BS 1362 fuses. It is perhaps sensible to include, at this point, circuit breakers to BS 3871, BS EN 60898.

Breaking capacity of fuses and circuit breakers

(IEE Reg 432–04–01)

When a short-circuit occurs, the current may, for a fraction of a second, reach hundreds or even thousands of amperes. The protective device must be able to break, and in the case of circuit breakers, make such a current without damage to its surroundings by arcing, overheating or the scattering of hot particles.

Circuit breakers	Breaking capacity (kA)	
BS3871 Types 1, 2, 3 etc.	1	(M1)
	1.5	(M1.5)
	3	(M3)
	4.5	(M4.5)
	6	(M6)
	9	(M9)
BS EN 60898 Types B, C, D	Icn	Ics
	1.5	1.5
	3	3
	6	6
	10	7.5
	15	7.5
	25	10

Icn is the rated ultimate breaking capacity.
Ics is the maximum breaking capacity operation after which the breaker may still be used without loss of performance.

The table on pages 61 and 64 indicate the performance of circuit breakers and the more commonly used British Standard fuse links.

Although all reference to BS 3871 MCB's have been removed from BS 7671, they are still used and therefore worthy of mention.

Fuse and circuit breaker operation

(IEE Regs 433-02 and 434-03)

Let us consider a protective device rated at, say, 10 A. This value of current can be carried indefinitely by the device, and is known as its nominal setting I_n. The value of the current which will cause operation of the device, I_2, will be larger than I_n, and will be dependent on the device's *fusing factor*. This is a figure which, when multiplied by the nominal setting I_n, will indicate the value of operating current I_2.

For fuses to BS 88 and BS 1361 and circuit breakers to BS 3871 this fusing factor is approximately 1.45; hence our 10 A device would not operate until the current reached $1.45 \times 10 = 14.5\,\text{A}$. The IEE Regulations require coordination between conductors and protection when an overload occurs, such that:

1 The nominal setting of the device I_n is greater than or equal to the design current of the circuit $I_b (I_n \geqslant I_b)$.
2 The nominal setting I_n is less than or equal to the lowest current-carrying capacity I_z of any of the circuit conductors $(I_n \leqslant I_z)$.
3 The operating current of the device I_2 is less than or equal to $1.45\, I_z (I_2 \leqslant 1.45\, I_z)$.

So, for our 10 A device, if the cable is rated at 10 A then condition 2 is satisfied. Since the fusing factor is 1.45, condition 3 is also satisfied: $I_2 = I_n \times 1.45 = 10 \times 1.45$, which is also 1.45 times the 10 A cable rating.

The problem arises when a BS 3036 semi-enclosed rewirable fuse is used, as it may have a fusing factor of as much as 2. In order to comply with condition 3, I_n should be less than or equal to $0.725\, I_z$. This figure is derived from $1.45/2 = 0.725$. For example, if a cable is rated at 10 A, then I_n for a BS 3036 should be $0.725 \times 10 = 7.25\,\text{A}$. As the fusing factor is 2, the operating current $I_2 = 2 \times 7.25 = 14.5$, which conforms with condition 3, i.e. $I_2 \leqslant 1.45 \times 120 = 14.5$.

All of these foregoing requirements ensure that conductor insulation is undamaged when an overload occurs.

Under short-circuit conditions it is the conductor itself that is susceptible to damage and must be protected. Figure 18 shows one half-cycle of short-circuit current if there were no protection. The RMS value $(0.7071 \times \text{maximum value})$ is called the prospective short-circuit current. The cut-off point is where the short-circuit current is interrupted and an arc is formed; the time t_1 taken to reach this point is called the pre-arcing time. After the current has been cut off, it falls to zero as the arc is being extinguished. The time t_2 is the total time taken to disconnect the fault.

During the time t_1, the protective device is allowing energy to pass through to the load side of the circuit. This energy is known as the pre-arcing let-through energy and is given by $I^2 t_1$, where I is the short-circuit current. The total let-through energy from start to disconnection of the fault is given by $I^2 t_2$ (see Figure 19).

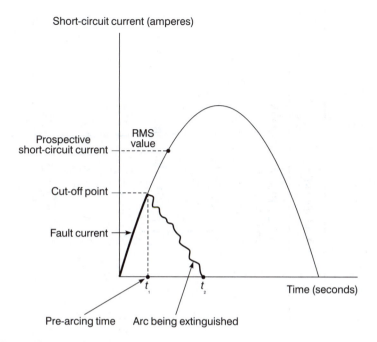

Figure 18

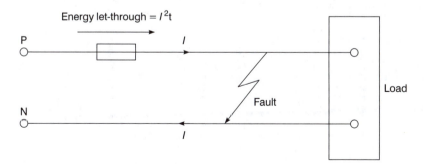

Figure 19

British Standards for fuse links

	Standard	Current rating	Voltage rating
1	BS 2950	Range 0.05–25 A	Range 1000 V (0.05 A) to 32 V (25 A) AC and DC
2	BS 646	1, 2, 3 and 5 A	Up to 250 V AC and DC
3	BS 1362 cartridge	1, 2, 3, 5, 7, 10 and 13 A	Up to 250 V AC
4	BS 1361 HRC cut-out fuses	5, 15, 20, 30, 45 and 60 A	Up to 250 V AC
5	BS 88 motors	Four ranges, 2–1200 A	Up to 660 V, but normally 250 or 415 V AC and 250 or 500 V DC
6	BS 2692	Main range from 5 to 200 A; 0.5 to 3 A for voltage transformer protective fuses	Range from 2.2 to 132 kV
7	BS 3036 rewirable	5, 15, 20, 30, 45, 60, 100, 150 and 200 A	Up to 250 V to earth
8	BS 4265	500 mA to 6.3 A 32 mA to 2 A	Up to 250 V AC

British Standards for fuse links (continued)

Breaking capacity	Notes
1 Two or three times current rating	Cartridge fuse links for telecommunication and light electrical apparatus. Very low breaking capacity
2 1000 A	Cartridge fuse intended for fused plugs and adapters to BS546: 'round-pin' plugs
3 6000 A	Cartridge fuse primarily intended for BS1363: 'flat pin' plugs
4 16 500 A 33 000 A	Cartridge fuse intended for use in domestic consumer units. The dimensions prevent interchangeability of fuse links which are not of the same current rating
5 Ranges from 10 000 to 80 000 A in four AC and three DC categories	Part 1 of Standard gives performance and dimensions of cartridge fuse links, whilst Part 2 gives performance and requirements of fuse carriers and fuse bases designed to accommodate fuse links complying with Part 1
6 Ranges from 25 to 750 MVA (main range) 50 to 2500 MVA (VT fuses)	Fuses for AC power circuits above 660 V
7 Ranges from 1000 to 12000 A	Semi-enclosed fuses (the element is a replacement wire) for AC and DC circuits
8 1500 A (high breaking capacity) 35 A (low breaking capacity)	Miniature fuse links for protection of appliances of up to to 250 V (metric standard)

For faults of up to 5 s duration, the amount of heat energy that cable can withstand is given by k^2s^2, where s is the cross-sectional area of the conductor and k is a factor dependent on the conduct or material. Hence, the let-through energy should not exceed k^2s^2, i.e. $I^2t = k^2s^2$. If we transpose this formula for t, we get $t = k^2s^2/I^2$, which is the maximum disconnection time in seconds.

Remember that these requirements refer to short-circuit currents only. If, in fact, the protective device has been selected to protect against overloads and has a breaking capacity not less than the prospective short-circuit current (PSCC) at the point of installation, it will also protect against short-circuit currents. However, if there is any doubt the formula should be used.

For example, in Figure 20, if I_n has been selected for overload protection, the questions to be asked are as follows:

1 Is $I_n \geqslant I_b$? Yes.
2 Is $I_n \leqslant I_z$? Yes.
3 Is $I_2 \geqslant 1.45I_z$? Yes.

Then, if the device has a rated breaking capacity not less than the PSCC, it can be considered to give protection against shortcircuit current also.

When an installation is being designed, the prospective shortcircuit current at every relevant point must be determined, by either calculation or measurement. The value will decrease as we move farther away from the intake position (resistance increases with

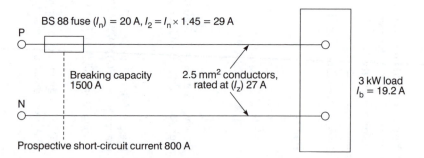

BS 88 fuse (I_n) = 20 A, $I_2 = I_n \times 1.45$ = 29 A

Breaking capacity 1500 A

2.5 mm^2 conductors, rated at (I_z) 27 A

3 kW load I_b = 19.2 A

Prospective short-circuit current 800 A

Figure 20

length). Thus, if the breaking capacity of the lowest rated fuse in the installation is greater than the prospective short-circuit at the origin of the supply, there is no need to determine the value except at the origin.

Discrimination

(IEE Reg 533-01-06)

When we discriminate, we indicate our preference over other choices: this house rather than that house, for example. With protection we have to ensure that the correct device operates when there is a fault. Hence, a 13 A BS 1362 plug fuse should operate before the main circuit fuse. Logically, protection starts at the origin of an installation with a large device and progresses down the chain with smaller and smaller sizes.

Simply because protective devices have different ratings, it cannot be assumed that discrimination is achieved. This is especially the case where a mixture of different types of device is used. However, as a general rule a 2:1 ratio with the lower-rated devices will be satisfactory. The table on page 57 shows how fuse links may be chosen to ensure discrimination.

Fuses will give discrimination if the figure in column 3 does not exceed the figure in column 2. Hence:

a 2 A fuse will discriminate with a 4 A fuse
a 4 A fuse will discriminate with a 6 A fuse
a 6 A fuse will *not* discriminate with a 10 A fuse
a 10 A fuse will discriminate with a 16 A fuse.

All other fuses will *not* discriminate with the next highest fuse, and in some cases, several sizes higher are needed e.g. a 250 A fuse will only discriminate with a 400 A fuse.

Position of protective devices

(IEE Regs Section 473)

When there is a reduction in the current-carrying capacity of a conductor, a protective device is required. There are, however,

I^2t characteristics: 2–800 A fuse links. Discrimination is achieved if the total I^2t of the minor fuse does not exceed the pre-arcing I^2t of the major fuse

Rating (A)	I^2t pre-arcing	I^2t total at 415 V
2	0.9	1.7
4	4	12
6	16	59
10	56	170
16	190	580
20	310	810
25	630	1 700
32	1 200	2 800
40	2 000	6 000
50	3 600	11 000
63	6 500	14 000
80	13 000	36 000
100	24 000	66 000
125	34 000	120 000
160	80 000	260 000
200	140 000	400 000
250	230 000	560 000
315	360 000	920 000
350	550 000	1 300 000
400	800 000	2 300 000
450	700 000	1 400 000
500	900 000	1 800 000
630	2 200 000	4 500 000
700	2 500 000	5 000 000
800	4 300 000	10 000 000

some exceptions to this requirement; these are listed quite clearly in Section 473 of the IEE Regulations. As an example, protection is not needed in a ceiling rose where the cable size changes from $1.0 \, \text{mm}^2$ to, say, $0.5 \, \text{mm}^2$ for the lampholder flex. This is permitted as it is not expected that lamps will cause overloads.

Protection against overvoltage

(IEE Regs Chapter 44)

This chapter deals with the requirements of an electrical installation to withstand overvoltages caused by lightning or switching surges. It is unlikely that installations in the UK will be affected by the requirements of this section as the number of thunderstorm days per year is not likely to exceed 25.

Protection against undervoltage

(IEE Regs Chapter 45 and Section 535)

From the point of view of danger in the event of a drop or loss of voltage, the protection should prevent automatic restarting of machinery etc. In fact, such protection is an integral part of motor starters in the form of the control circuit.

Precautions where there is a particular risk of fire

(IEE Regs Chapter 48)

This chapter outlines details of locations and situations where there may be a particular risk of fire. These would include locations where combustible materials are stored or could collect and where a risk of ignition exists. This section does not include locations where there is a risk of explosion (see page 133).

4
CONTROL

Definitions used in this chapter

Emergency switching Rapid cutting off of electrical energy to remove any hazard to persons, livestock or property which may occur unexpectedly.

Isolation Cutting off an electrical installation, a circuit, or an item of equipment from every source of electrical energy.

Mechanical maintenance The replacement, refurbishment or cleaning of lamps and non-electrical parts of equipment, plant and machinery.

Switch A mechanical switching device capable of making, carrying and breaking current under normal circuit conditions, which may include specified overload conditions, and also of carrying, for a specified time, currents under specified abnormal conditions such as those of short circuit.

Isolation and switching

(Relevant parts, sections and chapters Chapter 46 and Sections 476 and 537)

The essential part of a motor control circuit that will ensure under-voltage protection is the 'hold-on' circuit (Figure 21).

When the start button is pushed, the coil becomes energized and its normally open (N/O) contacts close. When the start button is

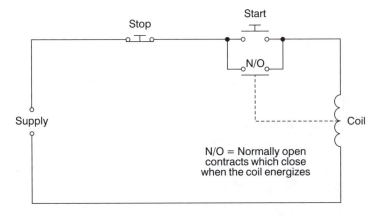

Figure 21   Hold-on circuit

released the coil remains energized via its own N/O contacts. These are known as the 'hold-on' contacts.

The coil can only be de-energized by opening the circuit by the use of the stop button or by a considerable reduction or loss of voltage. When this happens, the N/O contacts open and, even if the voltage is restored or the circuit is made complete again, the coil will remain de-energized until the start button is pushed again. Figure 22 shows how this 'hold-on' facility is built into a typical single-phase motor starter.

Having decided how we are going to earth an installation, and settled on the method of protecting persons and livestock from electric shock, and conductors and insulation from damage. We must now investigate the means of controlling the installation. In simple terms, this means the switching of the installation or any part of it 'on' or 'off'. The IEE Regulations refer to this topic as 'isolation and switching'.

By definition, isolation is the cutting off of electrical energy from every source of supply, and this function is performed by a switch, a switch fuse or a fuse switch. A switch is sometimes referred to as an isolator or a switch disconnector. A switch fuse is a switch housed in the same enclosure as the fuses. A fuse switch has the

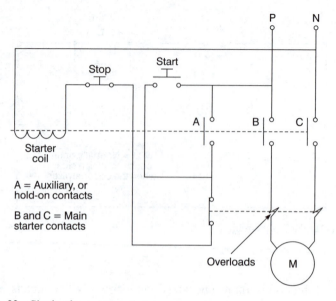

Figure 22 Single-phase motor starter

fuses as part of the switch mechanism, and is used mainly where higher load currents are anticipated.

On the grid system there is a distinct difference between an isolator and a switch. A switch, or breaker as it is known, is capable of breaking fault currents as well as normal load currents. By contrast, an isolator is installed for the purpose of isolating sections of the system for maintenance work etc. (see Figure 23). For work to be carried out on the breaker, the part of the network that the circuit supplies is fed from another source. The breaker is then operated and

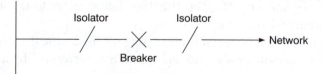

Figure 23

both isolators are opened, thus preventing supply from any source reaching the breaker.

With a domestic installation, the main switch in a consumer unit is considered to be a means of isolation for the whole installation, and each fuse or circuit breaker to be isolators for the individual circuits. Ideally all of these devices should have some means of preventing unintentional re-energization, either by locks or interlocks. In the case of fuses and circuit breakers, these can be removed and kept in a safe place.

In many cases, isolating and locking off come under the heading of switching off for mechanical maintenance. Hence, a switch controlling a motor circuit should have, especially if it is remote from the motor, a means of locking in the 'off' position (Figure 24).

A one-way switch controlling a lighting point is a functional switch, but could be considered as a means of isolation, or a means of switching off for mechanical maintenance (changing a lamp). A two-way switching system, however, does not provide a means of isolation, as neither switch cuts off electrical energy from all sources of supply.

In an industrial or workshop environment it is important to have a means of cutting off the supply to the whole or parts of the installation in the event of an emergency. The most common method is the provision of stop buttons suitably located and used in conjunction with a contactor or relay (Figure 25).

Pulling a plug from a socket to remove a hazard is not permitted as a means of emergency switching. It is, however, allowed as a

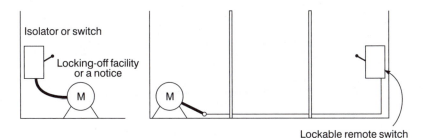

Isolator or switch

Locking-off facility or a notice

M

M

Lockable remote switch

Figure 24

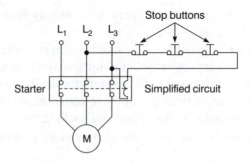

Figure 25

means of functional switching, e.g. switching off a hand lamp by unplugging.

Whilst we are on the subject of switching, it should be noted that a switch controlling discharge lighting (this includes fluorescent fittings) should, unless it is specially designed for the purpose, be capable of carrying at least twice the steady load of the circuit. The reason for this is that discharge lighting contains chokes which are inductive and cause arcing at switch contacts. The higher rating of the switch enables it to cope with arcing.

5
CIRCUIT DESIGN

Definitions used in this chapter

Ambient temperature   The temperature of the air or other medium where the equipment is to be used.

Circuit protective conductor   A protective conductor connecting exposed conductive parts of equipment to the main earthing terminal.

Current-carrying capacity   The maximum current which can be carried by a conductor under specified conditions without its steady state temperature exceeding a specified value.

Design current   The magnitude of the current intended to be carried by a circuit in normal service.

Earthing conductor   A protective conductor connecting a main earthing terminal of an installation to an earth electrode or other means of earthing.

Overcurrent   A current exceeding the rated value. For conductors the rated value is the current-carrying capacity.

Short-circuit current   An overcurrent resulting from a fault of negligible impedance between live conductors having a difference of potential under normal operating conditions.

Design procedure

(IEE Regs 514-09 and 712-01-03 (xviii);

IEE Regs Chapter 3; IEE Regs Appendix 4)

The requirements of IEE Regulations make it clear that circuits must be designed and the design data made readily available. In fact this has always been the case with previous editions of the Regulations, but it has not been so clearly indicated.

How then do we begin to design? Clearly, plunging into calculations of cable size is of little value unless the type of cable and its method of installation is known. This, in turn, will depend on the installation's environment. At the same time, we would need to know whether the supply was single- or three-phase, the type of earthing arrangements, and so on. Here then is our starting point and it is referred to in the Regulations, Chapter 3, as 'Assessment of general characteristics'.

Having ascertained all the necessary details, we can decide on an installation method, the type of cable, and how we will protect against electric shock and overcurrents. We would now be ready to begin the calculation part of the design procedure.

Basically there are eight stages in such a procedure. These are the same whatever the type of installation, be it a cooker circuit or a submain cable feeding a distribution board in a factory. Here, then, are the eight basic steps in a simplified form:

1 Determine the design current I_b.
2 Select the rating of the protection I_n.
3 Select the relevant correction factors (CFs).
4 Divide I_n by the relevant CFs to give tabulated cable current-carrying capacity I_t.
5 Choose a cable size, to suit I_t.
6 Check the voltage drop.
7 Check for shock risk constraints.
8 Check for thermal constraints.

Let us now examine each stage in detail.

Design current

In many instances the design current I_b is quoted by the manufacturer, but there are times when it has to be calculated. In that case there are two formulae involved, one for single-phase and one for three-phase:

Single-phase:

$$I_b = \frac{P \text{ (watts)}}{V} \quad (\text{V usually } 230\,\text{V})$$

Three phase:

$$I_b = \frac{P \text{ (watts)}}{\sqrt{3} \times V_L} \quad (V_L \text{ usually } 400\,\text{V})$$

Current is in amperes, and power P in watts.

If an item of equipment has a power factor (PF) and/or has moving parts, efficiency (eff) will have to be taken into account.

Hence:

Single-phase:

$$I_b = \frac{P \text{ (watts)} \times 100}{V \times PF \times eff}$$

Three-phase:

$$I_b = \frac{P \times 100}{\sqrt{3} \times V_L \times PF \times eff}$$

Nominal setting of protection

Having determined I_b we must now select the nominal setting of the protection such that $I_n \geqslant I_b$. This value may be taken from IEE Regulations, Tables 41B1, B2 or D or from manufacturers' charts. The choice of fuse or MCB or CB type is also important and may have to be changed if cable sizes or loop impedances are too high. These details will be discussed later.

Correction factors

When a cable carries its full load current it can become warm. This is no problem unless its temperature rises further due to other influences, in which case the insulation could be damaged by overheating. These other influences are: high ambient temperature; cables grouped together closely; uncleared overcurrents; and contact with thermal insulation.

For each of these conditions there is a correction factor (CF) which will respectively be called C_a, C_g, C_f and C_i, and which derates cable current-carrying capacity or conversely increases cable size.

Ambient temperature C_a

The cable ratings in the IEE Regulations are based on an ambient temperature of 30 °C, and hence it is only above this temperature that an adverse correction is needed. Table 4C 1 of the Regulations gives factors for all types of protection other than BS 3036 semi-enclosed rewirable fuses, which are accounted for in Table 4C2.

Grouping C_g

When cables are grouped together they impart heat to each other. Therefore, the more cables there are the more heat they will generate, thus increasing the temperature of each cable. Table 4B of the Regulations gives factors for such groups of cables or circuits. It should be noted that the figures given are for cables of the same size, and hence correction may not necessarily be needed for cables grouped at the outlet of a domestic consumer unit, for example, where there is a mixture of different sizes.

A typical situation where correction factors need to be applied would be in the calculation of cable sizes for a lighting system in a large factory. Here many cables of the same size and loading may be grouped together in trunking and could be expected to be fully loaded all at the same time.

Protection by BS 3036 fuse C_f

As we have already discussed in Chapter 3, because of the high fusing factor of BS 3036 fuses, the rating of the fuse I_n, should be $\leqslant 0.725\ I_z$. Hence 0.725 is the correction factor to be used when BS 3036 fuses are used.

Thermal insulation C_i

With the modern trend towards energy saving and the installation of thermal insulation, there may be a need to derate cables to account for heat retention.

The values of cable current-carrying capacity given in Appendix 4 of the IEE Regulations have been adjusted for situations when thermal insulation touches one side of a cable. However, if a cable is totally surrounded by thermal insulation for more than 0.5 m, a factor of 0.5 must be applied to the tabulated clipped direct ratings. For less than 0.5 m, derating factors (Table 52A of the Regulations) should be applied.

Application of correction factors

Some or all of the onerous conditions just outlined may affect a cable along its whole length or parts of it, but not all may affect it at the same time. So, consider the following:

1 If the cable in Figure 26 ran for the whole of its length, grouped with others of the same size in a high ambient temperature, and

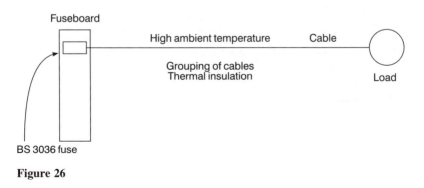

Figure 26

was totally surrounded with thermal insulation, it would seem logical to apply all the CFs, as they all affect the whole cable run. Certainly the factors for the BS 3036 fuse, grouping and thermal insulation should be used. However, it is doubtful if the ambient temperature will have any effect on the cable, as the thermal insulation, if it is efficient, will prevent heat reaching the cable. Hence, apply C_g, C_f and C_i.

2 In Figure 27(a) the cable first runs grouped, then leaves the group and runs in high ambient temperature, and finally is enclosed in thermal insulation. We therefore have three different conditions, each affecting the cable in different areas. The BS 3036 fuse affects the whole cable run and therefore C_f must be used, but there is no need to apply all of the remaining factors as the worse one will automatically compensate for the others. The relevant factors are shown in Figure 27(b): apply only $C_f = 0.725$ and

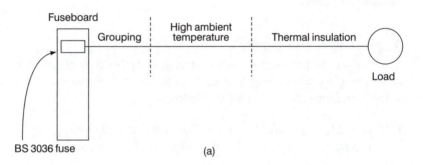

(a)

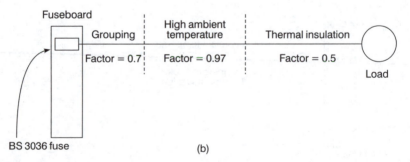

(b)

Figure 27

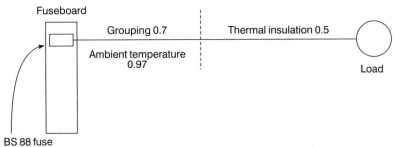

Figure 28

$C_i = 0.5$. If protection was *not* by BS 3036 fuse, then apply only $C_i = 0.5$.

3 In Figure 28 a combination of cases 1 and 2 is considered. The effect of grouping and ambient temperature is $0.7 \times 0.97 = 0.69$. The factor for thermal insulation is still worse than this combination, and therefore C_i is the only one to be used.

Having chosen the *relevant* correction factors, we now apply them to the nominal rating of the protection I_n as divisors in order to calculate the current-carrying capacity I_t of the cable.

Tabulated current-carrying capacity

The required formula for current-carrying capacity I_t is:

$$I_t \geqslant \frac{I_n}{\text{relevant CFs}}$$

In Figure 29 the current-carrying capacity is given by:

$$I_t \geqslant \frac{I_n}{C_f C_i} = \frac{30}{0.725 \times 0.5} = 82.75 \, \text{A}$$

or, without the BS 3036 fuse:

$$I_t \geqslant \frac{30}{0.5} = 60 \, \text{A}$$

Circuit design

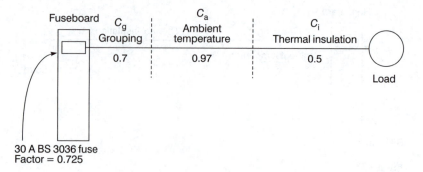

Figure 29

In Figure 30, $C_a C_g = 0.97 \times 0.5 = 0.485$, which is worse than C_i (0.5).

Hence:

$$I_t \geqslant \frac{I_n}{C_f C_a C_g} = \frac{30}{0.725 \times 0.485} = 85.3\,\text{A}$$

or, without the BS 3036 fuse:

$$I = \frac{30}{0.485} = 61.85\,\text{A}$$

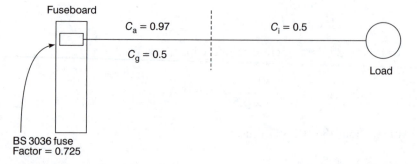

Figure 30

Note: If the circuit is not subject to overload, I_n can be replaced by I_b so the formula becomes:

$$I_t \geqslant \frac{I_b}{CFs}$$

Choice of cable size

Having established the tabulated current-carrying capacity I_t of the cable to be used, it now remains to choose a cable to suit that value. The tables in Appendix 4 of the IEE Regulations list all the cable sizes, current-carrying capacities and voltage drops of the various types of cable. For example, for PVC-insulated singles, single-phase, in conduit, having a current-carrying capacity of 45 A, the installation is by reference method 3 (Table 4A), the cable table is 4D1A and the column is 4. Hence, the cable size is $10.0 \, \text{mm}^2$ (column 1).

Voltage drop

(IEE Regs 525)

The resistance of a conductor increases as the length increases and/ or the cross-sectional area decreases. Associated with an increased resistance is a drop in voltage, which means that a load at the end of a long thin cable will not have the full supply voltage available (Figure 31).

The IEE Regulations require that the voltage drop V should not be so excessive that equipment does not function safely. They further indicate that a drop of no more than 4% of the nominal voltage at the *origin* of the circuit will satisfy. This means that:

1 For single-phase 230 V, the voltage drop should not exceed 4% of 230 V = 9.2 V.
2 For three-phase 400 V, the voltage drop should not exceed 4% of 400 V = 16 V.

Circuit design

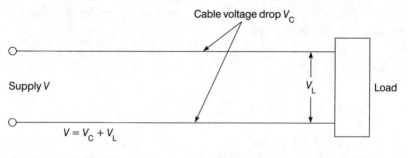

Figure 31

For example, the voltage drop on a circuit supplied from a 230 V source by a 16.0 mm² two core copper cable 23 m long, clipped direct and carrying a design current of 33 A, will be:

$$V_c = \frac{mV \times I_b \times L}{1000} \quad \text{(mV from Table 4D2B)}$$

$$= \frac{2.8 \times 33 \times 23}{1000} = 2.125 \, V$$

As we know that the maximum voltage drop in this instance (230 V) is 9.2 V, we can determine the maximum length by transposing the formula:

$$\text{maximum length} = \frac{V_c \times 1000}{mV \times I_b}$$

$$= \frac{9.2 \times 1000}{2.8 \times 23} = 142 \, m$$

There are other constraints, however, which may not permit such a length.

Shock risk

(IEE Regs 413-02-04)

This topic has already been discussed in full in Chapter 2. To recap, the actual loop impedance Z_s should not exceed those values given in Tables 41B1, B2 and D of the IEE Regulations. This ensures that circuits feeding socket outlets and equipment outside the equipotential zone will be disconnected, in the event of an earth fault, in less than 0.4 s, and that fixed equipment will be disconnected in less than 5 s.

Remember : $Z_s = Z_e + R_1 + R_2$.

Thermal constraints

(IEE Regs 543)

The IEE Regulations require that we either select or check the size of a CPC against Table 54F of the Regulations, or calculate its size using an adiabatic equation.

Selection of CPC using Table 54G

Table 54G of the Regulations simply tells us that:

1 For phase conductors up to and including $16\,mm^2$, the CPC should be at least the same size.
2 For sizes between $16\,m^2$ and $35\,mm^2$, the CPC should be at least $16\,mm^2$.
3 For sizes of phase conductor over $35\,mm^2$, the CPC should be at least half this size.

This is all very well, but for large sizes of phase conductor the CPC is also large and hence costly to supply and install. Also, composite cables such as the typical twin with CPC 6242Y type have CPCs smaller than the phase conductor and hence do not comply with Table 54G.

Calculation of CPC using an adiabatic equation

The adiabatic equation

$$s = \sqrt{(I^2 t)/k}$$

enables us to check on a selected size of cable, or on an actual size in a multicore cable. In order to apply the equation we need first to calculate the earth fault current from:

$$I = U_{oc}/Z_s$$

where U_{oc} is the transformer open circuit voltage (usually 240 V) and Z_s is the actual earth fault loop impedance. Next we select a k factor from Tables 54B to F of the Regulations, and then determine the disconnection time t from the relevant curve.

For those unfamiliar with such curves, using them may appear a daunting task. A brief explanation may help to dispel any fears. Referring to any of the curves for fuses in Appendix 3 of the IEE Regulations, we can see that the current scale goes from 1 A to 10 000 A, and the time scale from 0.01 s to 10 000 s. One can imagine the difficulty in drawing a scale between 1 A and 10 000 A in divisions of 1 A, and so a logarithmic scale is used. This cramps the large scale into a small area. All the subdivisions between the major divisions increase in equal amounts depending on the major division boundaries; for example, all the subdivisions between 100 and 1000 are in amounts of 100 (Figure 32).

Figures 34 and 35 give the IEE Regulations time/current curves for BS 88 fuses. Referring to the appropriate curve for a 32 A fuse (Figure 34), we find that a fault current of 200 A will cause disconnection of the supply in 0.6 s.

Where a value falls between two subdivisions, e.g. 150 A, an estimate of its position must be made. Remember that even if the scale is not visible, it would be cramped at one end; so 150 A would not fall half-way between 100 A and 200 A (Figure 33).

It will be noted in Appendix 3 of the Regulations that each set of curves is accompanied by a table which indicates the current that

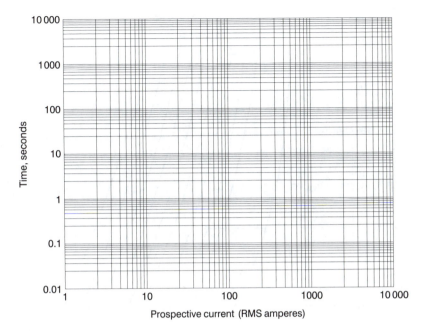

Figure 32

Figure 33

causes operation of the protective device for disconnection times of 0.1 s, 0.4 s and 5 s.

The IEE Regulations curves for CBs to BS EN 60898 type B and RCBO's are shown in Figure 36.

Having found a disconnection time, we can now apply the formula.

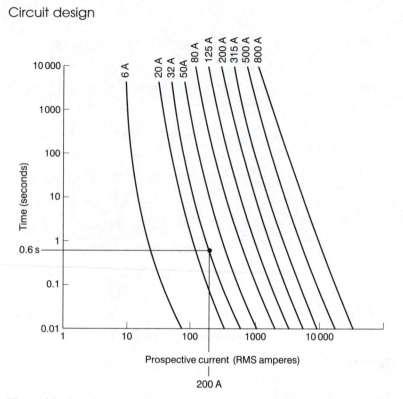

Figure 34 *Time/current characteristics for fuses to BS 88 Part 2. Example for 32 A fuse superimposed*

Example of use of an adiabatic equation

Suppose that in a design the protection was by 40 A BS 88 fuse; we had chosen a 4 mm² copper CPC running with our phase conductor; and the loop impedance Z_s was 1.2 Ω Would the chosen CPC size be large enough to withstand damage in the event of an earth fault? We have:

$$I = U_{oc}/Z_s = 240/1.2 = 200\,\text{A}$$

From the appropriate curve for the 40 A BS 88 fuse (Figure 35), we obtain a disconnection time t of 2 s. From Table 54C of the

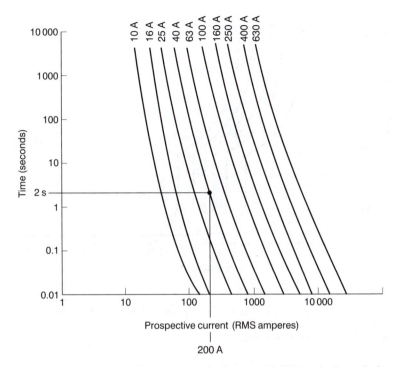

Figure 35 *Time/current characteristics for fuses to BS 88 Part 2. Example for 40 A fuse superimposed*

Regulations, $k = 115$. Therefore the minimum size of CPC is given by:

$$s = \sqrt{(I^2 t)}/k = \sqrt{200^2 \times 2}/115 = 2.46\,\text{mm}^2$$

So our $4\,\text{mm}^2$ CPC is acceptable. Beware of thinking that the answer means that we could change the $4\,\text{mm}^2$ for a $2.5\,\text{mm}^2$. If we did, the loop impedance would be different and hence I and t would change; the answer for s would probably tell us to use a $4\,\text{mm}^2$. In the example shown, s is merely a check on the actual size chosen.

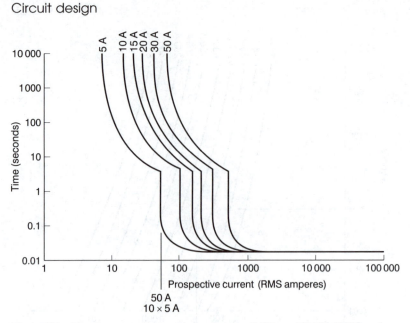

Figure 36 *Time/current characteristics for type 3 CBs to BS EN 60898 and RCBO's. Example for 50 A superimposed. For times less than 20 ms, the manufacturer should be consulted*

Having discussed each component of the design procedure, we can now put all eight together to form a complete design.

Example of circuit design

A consumer lives in a bungalow with a detached garage and workshop, as shown in Figure 37. The building method is traditional brick and timber.

The mains intake position is at high level, and comprises an 80 A BS 1361 240 V main fuse, an 80 A rated meter and a six-way 80 A consumer unit housing BS 3036 fuses as follows:

Ring circuit	30 A
Lighting circuit	5 A
Immersion heater circuit	15 A

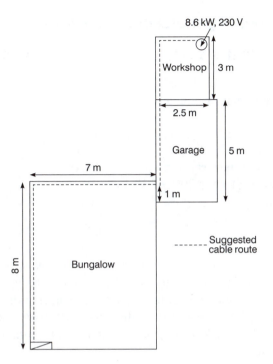

8.6 kW, 230 V

Workshop 3 m

2.5 m

Garage 5 m

7 m

1 m

Suggested cable route

8 m

Bungalow

Figure 37

Cooker circuit 30 A
Shower circuit 30 A
Spare way

The cooker is 40 A, with no socket in the cooker unit.

The main tails are 16 mm² double-insulated PVC, with a 6 mm² earthing conductor. There is no main equipotential bonding. The earthing system is TN–S, with an external loop impedance Z_e of 0.3 Ω. The prospective short-circuit current (PSCC) at the origin of the installation has been measured at 800 A. The roof space is insulated to the full depth of the ceiling joists, and the temperature in the roof space has been noted to be no more than 40 °C.

The consumer wishes to convert the workshop into a pottery room and install an 8.6 kW, 230 V electric kiln. The design procedure is as follows.

Assessment of general characteristics

(IEE Regs Part 3 & Appendix 3 on-site guide)

Present maximum demand

Applying diversity, we have:

Ring	30 A
Lighting (66% of 5 A)	3.3 A
Immersion heater	15 A
A cooker (10 A + 30% of 30 A)	19 A
Shower	30 A
Total	97.3 A

Reference to Table 4D1A of the Regulations will show that the existing main tails are too small and should be uprated. Also, the consumer unit should be capable of carrying the full load of the installation *without* the application of diversity. So the addition of another 8.6 kW of load is not possible with the present arrangement.

New maximum demand

The current taken by the kiln is $8600/230 = 37.4$ A. Therefore the new maximum demand is $127.3 + 37.4 = 164.7$ A.

Supply details

Single phase
230 V, 50 Hz
Earthing: TN–S
PFC at origin (measured between live conductors): 800 A

Decisions must now be made as to the type of cable, the installation method and the type of protective device. As the existing arrangement is not satisfactory, the supply authority must be informed of the new maximum demand, as a larger main fuse and service cable

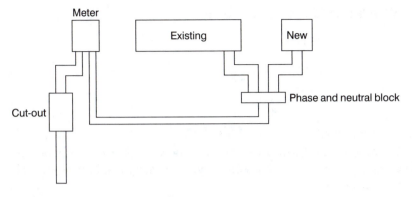

Figure 38

may be required. It would then seem sensible to disconnect, say, the shower circuit, and to supply it and the new kiln circuit via a new two-way consumer unit, as shown in Figure 38.

Sizing the main tails

(IEE Reg 547–02)

1 The new load on the existing consumer unit will be the old load less the shower load: $97.3 - 30 = 67.3$ A. From Table 4D1A of the Regulations, the cable size is $16\,\text{mm}^2$.
2 The load on the new consumer unit will be the kiln load plus the shower load: $37.4 + 30 = 67.4$ A. From Table 4D1A, the cable size is $16\,\text{mm}^2$.
3 The total load is $67.3 + 67.4 = 134.7$ A. From Table 4D1A, the cable size is $35\,\text{mm}^2$.
4 The earthing conductor size, from Table 54G, will be $16\,\text{mm}^2$. The main equipotential bonding conductor size, from Regulation 547–02, will be $10\,\text{mm}^2$.

For a domestic installation such as this, a PVC flat twin cable clipped direct through the loft space and the garage etc. would be most appropriate.

93

Sizing the kiln circuit cable

Design current

$$I_b = \frac{P}{V} = \frac{8600}{230} = 37.4\,\text{A}$$

Rating and type of protection

In order to show how important this choice is, it is probably best to compare the values of current-carrying capacity resulting from each type of protection.

As we have seen, the requirement for the rating I_n is that $I_n \geqslant I_b$. Therefore, using Table 41B2, I_n will be as follows for the various fuse types:

BS 88	40 A	BS 3036	45 A
BS 1361	45 A	MCB	50 A

Correction factors

C_a 0.87 or 0.94 if fuse is BS 3036
C_g not applicable
C_f 0.725 only if fuse is BS 3036
C_i 0.5 if cable is totally surrounded in thermal insulation

Tabulated current-carrying capacity of cable

For each of the different types of protection, the current-carrying capacity I_t will be as shown in Table (a), page 94.

Cable size based on tabulated current-carrying capacity

Table (b) on page 83 shows the sizes of cable for each type of protection (taken from Table 4D2 of the IEE Regulations).

Clearly the BS 88 fuse gives the smallest cable size if the cable is kept clear of thermal insulation, i.e. $6.0\,\text{mm}^2$.

(a)

	BS 88 40A	BS 1361 45 A	BS 3036 45 A	MCB 50 A
Surrounded by thermal insulation	$\dfrac{40}{0.5 \times 0.87} = 92$ A	$\dfrac{45}{0.5 \times 0.87} = 103.4$ A	$\dfrac{45}{0.5 \times 0.94 \times 0.725} = 132$ A	$\dfrac{50}{0.95 \times 0.87} = 115$ A
Not touching	$\dfrac{40}{0.87} = 46$ A	$\dfrac{45}{0.87} = 51.7$ A	$\dfrac{45}{0.94 \times 0.725} = 66$ A	$\dfrac{50}{0.87} = 57.5$ A

(b)

	BS 88	BS 1361	BS 3036	MCB
Cable size with thermal insulation	25.0 mm²	25.0 mm²	35.0 mm²	35.0 mm²
Cable size without thermal insulation	6.0 mm²	10.0 mm²	16.0 mm²	10.0 mm²
Cable size with half thermal insulation*	10.0 mm²	16.0 mm²	25.0 mm²	25.0 mm²

* See item number 15. Table 4A IEE Regulations.
In method 4, correction has already been made for cables touching thermal insulation on one size only.

Check on voltage drop

The actual voltage drop is given by:

$$\frac{mV \times I_b \times L}{1000} = \frac{7.3 \times 37.4 \times 24.5}{1000} = 6.7\,V$$

This voltage drop, whilst not causing the kiln to work unsafely, may mean inefficiency, and it is perhaps better to use a $10\,mm^2$ cable. This also gives us a wider choice of protection type, except BS 3036 rewirable. This decision we can leave until later.

For a $10\,mm^2$ cable, the voltage drop is checked as:

$$\frac{4.4 \times 37.4 \times 24.5}{1000} = 4.04\,V$$

So at this point we have selected a $10\,mm^2$ twin cable. We have at our disposal a range of protection types, the choice of which will be influenced by the loop impedance.

Shock risk

The CPC inside a $10\,mm^2$ twin 6242Y cable is $4\,mm^2$. Hence, the total loop impedance will be:

$$Z_s = Z_e + R_1 + R_2$$

For our selected cable, $R_1 + R_2$ for $24.5\,m$ will be (from tables of conductor resistance):

$$\frac{6.44 \times 1.2 \times 24.5}{1000} = 0.189\,\Omega$$

Note: the multiplier 1.2 takes account of the conductor resistance at its operating temperature.

We are given that $Z_e = 0.3\,\Omega$. Hence:

$$Z_s = 0.3 + 0.189 = 0.489\,\Omega$$

This means that all but 50 A types C and D CBs could be used (by comparison of values in Table 41 B1 and 41 B2 of the Regulations). As only BS EN 60898 types B, C and D will be available in the future, we must use type B.

Thermal constraints

We still need to check that the $4\,\text{mm}^2$ CPC is large enough to withstand damage under earth fault conditions. We have:

$$I = \frac{U_{oc}}{Z_s} = \frac{240}{0.489} = 490\,\text{A}$$

The disconnection time t for each type of protection from the relevant curves in the IEE Regulations are as follows:

40 A	BS 88	0.05 s
45 A	BS 1361	0.18 s
50 A	CB type B	0.01 s

From Table 54C of the Regulations, $k = 115$. Now:

$$s = \sqrt{(I^2 t)}/k$$

Therefore for each type of protection we have the following sizes s:

40 A	BS 88	$0.9\,\text{mm}^2$
45 A	BS 1361	$1.7\,\text{mm}^2$
50 A	CB type B	$0.466\,\text{mm}^2$

Hence, our $4\,\text{mm}^2$ CPC is of adequate size.

Protection

It simply remains to decide on the type of protection. Probably a type B CB is the most economical. However, if this is chosen a check should be made on the shower circuit to ensure that this type of protection is also suitable.

Design problem

In a factory it is required to install, side by side, two three-phase 400 V direct on line motors, each rated at 19 A full load current. There is spare capacity in a three-phase distribution fuseboard housing BS 3036 fuses, and the increased load will not affect the existing installation. The cables are to be PVC-insulated singles installed in steel conduit, and a separate CPC is required (note Regulation 543–1–02). The earthing system is TN–S with a measured external loop impedance of 0.47 Ω, and the length of the cable run is 42 m. The worst conduit section is 7 m long with one bend. The ambient temperature is not expected to exceed 40 °C.

Determine the minimum sizes of cable and conduit.

6
INSPECTION
AND TESTING

Definitions used in this chapter

Earth electrode   A conductor or group of conductors in intimate contact with and providing an electrical connection with earth.

Earth fault loop impedance   The impedance of the earth fault loop (phase-to-earth loop) starting and ending at the point of earth fault.

Residual current device   An electromechanical switching device or association of devices intended to cause the opening of the contacts when the residual current attains a given value under specified conditions.

Ring final circuit   A final circuit arranged in the form of a ring and connected to a single point of supply.

Testing sequence (Part 7)

Having designed our installation, selected the appropriate materials and equipment, and installed the system, it now remains to put it into service. However, before the installation is put into service it must be tested and inspected to ensure that it complies, as far as is practicable, with the IEE Regulations. Note the word 'practicable';

it would be unreasonable, for example, to expect the whole length of a circuit cable to be inspected for defects, as this may mean lifting floorboards etc.

Part 7 of the IEE Regulations give details of testing and inspection requirements. Unfortunately, these requirements presuppose that the person carrying out the testing is in possession of all the design data, which is only likely to be the case on the larger commercial or industrial projects. It may be wise for the person who will eventually sign the test certificate to indicate that the test and inspection were carried out as far as was possible in the absence of any design or other information.

However, let us continue by examining the required procedures. The Regulations initially call for a visual inspection, but some items such as correct connection of conductors etc. can be done during the actual testing. A preferred sequence of tests is recommended, where relevant, and is as follows:

1 Continuity of protective conductors.
2 Continuity of ring final circuit conductors.
3 Insulation resistance.
4 Insulation of site-built assemblies.
5 Protection by electrical separation.
6 Protection by barriers and enclosures provided during erection.
7 Insulation of non-conducting floors and walls.
8 Polarity.
9 Earth electrode resistance.
10 Earth fault loop impedance.
11 Prospective fault current.
12 Functional testing.

Not all of the tests may be relevant, of course. For example, in a domestic installation (TN–S or TN–C–S) only tests 1, 2, 3, 8, 9 and 11 would be needed.

The Regulations indicate quite clearly the tests required in Part 7. Let us then take a closer look at some of them in order to understand the reasoning behind them.

Continuity of protective conductors

All protective conductors, including main equipotential and supplementary bonding conductors, must be tested for continuity using a low-reading ohmmeter.

For main equipotential bonding there is no single fixed value of resistance above which the conductor would be deemed unsuitable. Each measured value, if indeed it is measurable for very short lengths, should be compared with the relevant value for a particular conductor length and size. Such values are shown in Table 3.

Where a supplementary equipotential bonding conductor has been installed between *simultaneously accessible* exposed and extraneous conductive parts, because circuit disconnection times cannot be met, then the resistance R of the conductor must be equal to or less than $50/I_a$. So:

$$R \leqslant 50/I_a.$$

Where 50 is the voltage, above which exposed metalwork should not rise, and I_a is the minimum current, causing operation of the circuit protective device within 5 s.

For example, suppose a 45 A BS3036 fuse protects a cooker circuit. The disconnection time for the circuit cannot be met, and

Table 3

Conductor size (mm²)	Resistance (mΩ/m)
1.0	18.1
1.5	12.1
2.5	7.41
4.0	4.61
6.0	3.08
10.0	1.83
16.0	1.15
25.0	0.727
35.0	0.524

so a supplementary bonding conductor has been installed between the cooker case and the adjacent metal sink. The resistance R of that conductor should not be greater than $50/I_a$ which in this case is 145 A (IEE Regulations). So:

$$50/145 = 0.34 \, \Omega.$$

How then, do we conduct a test to establish continuity of main or supplementary bonding conductors? Quite simple really: just connect the leads from the continuity tester to the ends of the bonding conductor (Figure 39). One end should be disconnected from its bonding clamp, otherwise any measurement may include the resistance of parallel paths of other earthed metalwork. Remember to zero or null the instrument first or, if this facility is not available, record the resistance of the test leads so that this value can be subtracted from the test reading.

IMPORTANT NOTE: If the installation is in operation, then never disconnect main bonding conductors unless the supply can be isolated. Without isolation, persons and livestock are at risk of electric shock.

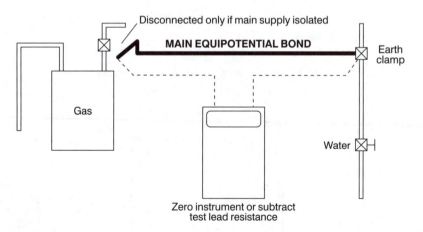

Figure 39

The continuity of circuit protective conductors may be established in the same way, but a second method is preferred, as the results of this second test indicate the value of $(R_1 + R_2)$ for the circuit in question.

The test is conducted in the following manner:

1 Temporarily link together the phase conductor and CPC of the circuit concerned in the distribution board or consumer unit.
2 Test between phase and CPC at each outlet in the circuit. A reading indicates continuity.
3 Record the test result obtained at the furthest point in the circuit. This value is $(R_1 + R_2)$ for the circuit.

Figure 40 illustrates the above method.

There may be some difficulty in determining the $(R_1 + R_2)$ values of circuits in installations that comprise steel conduit and trunking, and/or SWA and mims cables because of the parallel earth paths that are likely to exist. In these cases, continuity tests may have to be carried out at the installation stage before accessories are connected or terminations made off as well as after completion.

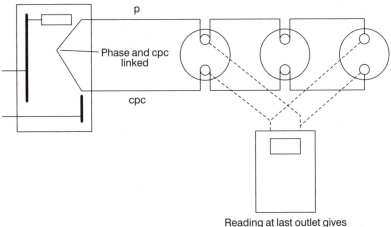

Reading at last outlet gives
the value of (R1 + R2) for the circuit

Figure 40

Continuity of ring final circuit conductors

There are two main reasons for conducting this test:

1 To establish that interconnections in the ring do not exist.
2 To ensure that the CPC is continuous, and indicate the value of $(R_1 + R_2)$ for the ring.

What then are interconnections in a ring circuit, and why is it important to locate them? Figure 41 shows a ring final circuit with an interconnection.

The most likely cause of the situation shown in Figure 41 is where a DIY enthusiast has added sockets P, Q, R and S to an existing ring A, B, C, D, E and F. In itself there is nothing wrong with this. The problem arises if a break occurs at, say, point Y, or the terminations fail in socket C or P. Then there would be four sockets all fed from the point X which would then become a spur. So, how do we identify such a situation with or without breaks at point Y? A simple resistance test between the ends of the phase, neutral or circuit protective conductors will only indicate that a circuit exists, whether there are interconnections or not. The following test method based on the theory that the resistance measured across any diameter of a perfect circle of conductor will always be the same value (Figure 42).

The perfect circle of conductor is achieved by cross-connecting the phase and neutral loops of the ring (Figure 43).

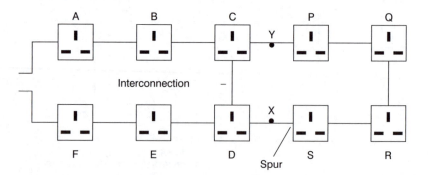

Figure 41

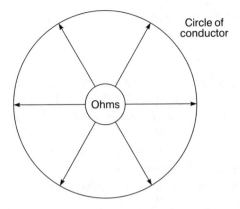

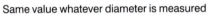

Same value whatever diameter is measured

Figure 42

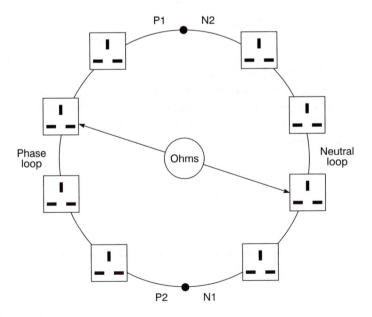

Figure 43

The test procedure is as follows:

1 *Identify the opposite legs of the ring.* This is quite easy with sheathed cables, but with singles, each conductor will have to be identified, probably by taking resistance measurements between each one and the closest socket outlet. This will give three high readings and three low readings, thus establishing the opposite legs.
2 *Take a resistance measurement between the ends of each conductor loop. Record this value.*
3 *Cross connect the ends of the phase and neutral loops (see Figure 44).*
4 *Measure between phase and neutral at each socket on the ring.* The readings obtained should be, for a perfect ring, substantially the same.

If an interconnection existed such as shown in Figure 41 then sockets A to F would all have similar readings, and those beyond the interconnection would have gradually increasing values to approximately the midpoint of the ring, then decreasing values back towards the interconnection. If a break had occurred at point Y then the readings from socket S would increase to a maximum at socket P. One or two high readings are likely to indicate either loose connections or spurs. A null reading, i.e. an open circuit indication, is probably a reverse polarity, either phase-CPC or neutral-CPC reversal. These faults would clearly be rectified and the test at the suspect socket(s) repeated.

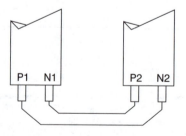

Figure 44

5 *Repeat the above procedure, but in this case cross connect the phase and CPC loops.* In this instance, if the cable is of the flat twin type, the readings at each socket will very slightly increase and then decrease around the ring. This difference, due to the phase and CPC being different sizes, will not be significant enough to cause any concern. The measured value is very important, it is $R_1 + R_2$ for the ring.

As before, loose connections, spurs and, in this case P–N cross-polarity, will be picked up.

The details that follow are typical approximate ohmic values for a healthy 70 m ring final circuit wired in 2.5/1.5 flat twin and CPC cable:

	P1 to P2	N1 to N2	CPC1 to CPC2
Initial measurements: Reading at each socket:	0.26	0.26	between 0.32 and 0.34
For spurs, each metre in length will add the following resistance to the above values:	0.015	0.015	0.02

Insulation resistance

This is probably the most used and yet most abused test of them all. Affectionately known as 'meggering', an insulation resistance test is performed in order to ensure that the insulation of conductors, accessories and equipment is in a healthy condition, and will prevent dangerous leakage currents between conductors and between conductors and earth. It also indicates whether any short circuits exist.

Insulation resistance is the resistance measured between conductors and is made up of countless millions of resistances in parallel (Figure 45).

The more resistances there are in parallel, the lower the overall resistance, and in consequence, the longer a cable the lower the insulation resistance. Add to this the fact that almost all installation circuits are also wired in parallel, and it becomes apparent

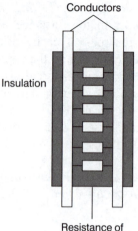

Figure 45

that tests on large installations may give, if measured as a whole, pessimistically low values, even if there are no faults. Under these circumstances, it is usual to break down such large installations into smaller sections, floor by floor, sub-main by sub-main etc. This also helps, in the case of periodic testing, to minimize disruption.

The test procedure then, is as follows:

1 Disconnect all items of equipment such as capacitors and indicator lamps as these are likely to give misleading results. Remove any items of equipment likely to be damaged by the test, such as dimmer switches, electronic timers etc. Remove all lamps and accessories and disconnect fluorescent and discharge fittings. Ensure that the installation is disconnected from the supply, all fuses are in place, and MCBs and switches are in the on position. In some instances it may be impracticable to remove lamps etc. and in this case the local switch controlling such equipment may be left in the off position.

2 Join together all live conductors of the supply and test between this join and earth. Alternatively, test between each live conductor and earth in turn.

3 Test between phase and neutral. For three-phase systems, join together all phases and test between this join and neutral. Then test between each of the phases. Alternatively, test between each of the live conductors in turn. Installations incorporating two-way lighting systems should be tested twice with the two-way switches in alternative positions.

The following Table 4 gives the test voltages and minimum values of insulation resistance for ELV and LV systems.

If a value of less than $2\,\text{M}\Omega$ is recorded it may indicate a situation where a fault is developing, but as yet still complies with the minimum permissible value. In this case each circuit should be tested separately and each should be above $2\,\text{M}\Omega$.

Polarity

This simple test, often overlooked, is just as important as all the others, and many serious injuries and electrocutions could have been prevented if only polarity checks had been carried out.

The requirements are:

- all fuses and single pole switches are in the phase conductor
- the centre contact of an Edison screw type lampholder is connected to the phase conductor
- all socket outlets and similar accessories are correctly wired.

Although polarity is towards the end of the recommended test sequence, it would seem sensible, on lighting circuits, for example,

Table 4

System	Test voltage	Minimum insulation resistance
SELV and PELV	250 V DC	$0.25\,\text{M}\Omega$
LV up to 500 V	500 V DC	$0.5\,\text{M}\Omega$
Over 500 V	1000 V DC	$1.0\,\text{M}\Omega$

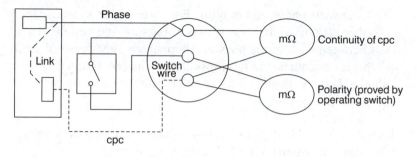

Figure 46

to conduct this test at the same time as that for continuity of CPCs, Figure 46.

As discussed earlier, polarity on ring final circuit conductors is achieved simply by conducting the ring circuit test. For radial socket outlet circuits, however, this is a little more difficult. The continuity of the CPC will have already been proved by linking phase and CPC and measuring between the same terminals at each socket. Whilst a phase-CPC reversal would not have shown, a phase-neutral reversal would, as there would have been no reading registered at the socket in question. This would have been remedied, and so only phase-CPC reversals need to be checked. This can be done by linking together phase and neutral at the origin and testing between the same terminals at each socket. A phase-CPC reversal will result in no reading at the socket in question.

Earth electrode resistance

As we know, in many rural areas, the supply system is T–T and hence reliance is placed on the general mass of earth for a return path under earth fault conditions and the connection to earth is made by an electrode, usually of the rod type.

In order to determine the resistance of the earth return path, it is necessary to measure the resistance that the electrode has with earth. In this instance an earth fault loop impedance test is carried out between the incoming phase terminal and the electrode (a standard test for Z_e). The value obtained is added to the CPC resistance of

Table 5 Values of loop impedance for comparison with test readings

RATING OF PROTECTION

Protection	Disconnection time		5A	6A	10A	15A	16A	20A	25A	30A	32A	40A	45A	50A	60A	63A	80A	100A	125A	160A	200A
BS 3036 fuse	0.4 s	Zs max.	7.5	//////	//////	2	//////	1.38	//////	0.85	//////	//////	0.46								
	5 s	Zs max.	13.9	//////	//////	4.18	//////	3	//////	2.97	//////	//////	1.25	//////	0.87			0.42			
BS 88 fuse	0.4 s	Zs max.	//////	6.66	4	//////	2.11	1.38	1.12	//////	0.82	0.64	//////	0.47							
	5 s	Zs max.	//////	10.5	5.8	//////	3.27	2.28	1.8	//////	1.44	1.05	//////	0.82	//////H	0.64	0.45	0.33	0.26	0.2	0.14
BS 1361 fuse	0.4 s	Zs max.	8.17	//////	//////	2.57	//////	1.33	//////	0.9	//////	//////	0.45								
	5 s	Zs max.	12.8	//////	//////	3.9	//////	2.19	//////	1.44	//////	//////	0.75	//////H	0.54	//////	0.39	0.28			
BS 3871 MCB Type 1	0.4 & 5 s	Zs max.	9	7.5	4.5	3	2.81	2.25	1.8	1.5	1.41	1.12	1	0.9	//////	0.71					
BS 3871 MCB Type 2	0.4 & 5 s	Zs max.	5.14	4.28	2.57	1.71	1.6	1.28	1.02	0.85	0.8	0.64	0.57	0.51	//////	0.4					
BS 3871 MCB Type 3	0.4 & 5 s	Zs max.	3.6	3	1.8	1.2	1.12	0.9	0.72	0.6	0.56	0.45	0.4	0.36	//////	0.28					
BS EN 60898 MCB Type B	0.4 & 5 s	Zs max.	//////	6	3.6	//////	2.25	1.8	1.44	//////	1.12	0.9	0.8	0.72	//////	0.57					
BS EN 60898 MCB Type C	0.4 & 5 s	Zs max.	3.6	3	1.8	1.2	1.12	0.9	0.72	0.6	0.56	0.45	0.4	0.36	//////	0.28					
BS EN 60898 MCB Type D	0.4 & 5 s	Zs max.	1.8	1.5	0.9	0.6	0.56	0.45	0.36	0.3	0.28	0.22	0.2	0.18	//////	0.14					

the protected circuits and this value is multiplied by the operating current of the RCD. The resulting value should not exceed 50 V.

Earth fault loop impedance

Overcurrent protective devices must, under earth fault conditions, disconnect fast enough to reduce the risk of electric shock. This is achieved if the actual value of the earth fault loop impedance does not exceed the tabulated maximum values given in BS 7671. The purpose of the test, therefore, is to determine the actual value of the loop impedance Z_s, for comparison with those maximum values, and it is conducted as follows:

1 Ensure that all main equipotential bonding is in place.
2 Connect the test instrument either by its BS4363 plug, or the 'flying leads', to the phase, neutral and earth terminals at the remote end of the circuit under test. (If a neutral is not available, connect the neutral probe to earth.)
3 Press to test and record the value indicated.

It must be understood that this instrument reading is *not valid for direct comparison with the tabulated maximum values*, as account must be taken of the ambient temperature at the time of test, and the maximum conductor operating temperature, both of which will have an effect on conductor resistance. Hence, the $(R_1 + R_2)$ is likely to be greater at the time of fault than at the time of test.

So, our measured value of Z_s must be corrected using correction factors and applying them in a formula.

Clearly this method of correcting Z_s is time-consuming and unlikely to be commonly used. Hence, a rule of thumb method may be applied which simply requires that the measured value of Z_s does not exceed 3/4 of the appropriate tabulated value. Table 7 gives the 3/4 values of tabulated loop impedance for direct comparison with measured values.

In effect, a loop impedance test places a phase/earth fault on the installation, and if an RCD is present it may not be possible to conduct the test, as the device will trip out each time the loop impedance tester button is pressed. Unless the instrument is of a type that has a built-in guard against such tripping, the value of Z_s will have to be determined from measured values of Z_e and $(R_1 + R_2)$.

NEVER SHORT OUT AN RCD IN ORDER TO CONDUCT THIS TEST

As a loop impedance test creates a high earth fault current, albeit for a short space of time, some lower rated MCBs may operate resulting in the same situation as with an RCD, and Z_s will have to be calculated. It is not really good practice to temporarily replace the MCB with one of a higher rating.

External loop impedance Z_e

The value of Z_e is measured at the intake position on the supply side and with all main equipotential bonding disconnected. Unless the installation can be isolated from the supply, this test should not be carried out, as a potential shock risk will exist with the supply on and the main bonding disconnected.

Prospective fault current

Prospective fault current (PFC) has to be determined at the origin of the installation. This is achieved by enquiry, calculation or measurement.

Functional testing

RCD RCBO operation

Where RCDs RCBOs are fitted, it is essential that they operate within set parameters. The RCD testers used are designed to do just this, and the basic tests required are as follows:

1 Set the test instrument to the rating of the RCD.
2 Set the test instrument to half rated trip.

3 Operate the instrument and the RCD should not trip.
4 Set the instrument to deliver the full rated tripping current of the RCD.
5 Operate the instrument and the RCD should trip out in the required time.

Table 6 gives further details.

Table 6

RCD type	Half rated	Full trip current
BS 4293 and BS 71–88 sockets	no trip	less than 200 ms
BS 4293 with time delay	no trip	$\frac{1}{2}$ time delay + 200 ms
BS EN 61009 or BS EN 61009 RCBO	no trip	300 ms
As above, but type S with time delay	no trip	130–500 ms

When an RCD is used for supplementary protection against direct contact, it must be rated at 30 mA or less and operate within 40 ms when subjected to a tripping current of five times its operating current.

Test instruments have the facility to provide this value of tripping current. There is no point in conducting this so-called fast trip test if an RCD has a rating in excess of 30 mA.

All RCDs have a built-in test facility in the form of a test button. Operating this test facility creates an artificial out of balance condition that causes the device to trip. This only checks the mechanics of the tripping operation, it is not a substitute for the tests just discussed.

All other items of equipment such as switchgear, control-gear interlocks etc. must be checked to ensure that they are correctly mounted and adjusted and that they function correctly.

7
SPECIAL LOCATIONS PART 6 BS 7671 (INCLUDING CHAPTER 48)

Introduction

The bulk of BS 7671 relates to typical, single and three phase, installations. There are, however, some locations within an installation that require special consideration. Such locations may present the user/occupant with an increased risk of death or injuries from electric shock.

BS 7671 categorises these special locations in Part 6 and they comprise installations in the following:

Section 601	Bathrooms, shower rooms etc.
Section 602	Swimming pools, indoor and outdoor
Section 603	Electrically heated saunas
Section 604	Construction sites
Section 605	Agricultural/horticultural locations
Section 606	Restrictive conductive locations
Section 607	Installations with high protective conductor currents
Section 608	Caravans and caravan sites
Section 611	Highway power supplies and street furniture.

Let us now briefly investigate the main requirements for each of these special locations.

Section 601: Bathrooms etc.

This section deals with rooms containing baths, shower basins or areas where showers exist but with tiled floors, e.g. Leisure/recreational centres, sports complexes etc.

Each of these areas are divided into zones which give an indication of the equipment/wiring etc. that can be installed in order to reduce the risk of electric shock.

Figure 47 indicating these zones, will probably be familiar to many readers.

So! out with the tape measure, only to find that in a one-bedroom flat, there may be no zone 2 or zone 3 etc. How can you conform to BS 7671?

The stark answer (mine) is that you may not be able to conform exactly. You do the very best you can in each particular circumstance to ensure safety. Let us not forget that the requirements of BS 7671 are based on reasonableness.

Hence, what electrical equipment can be installed in the zones (bonding comes later)?

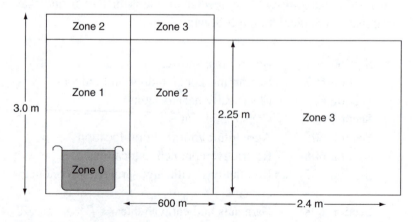

Figure 47 Bath/shower room zones

To start with, what equipment are we likely to encounter in the modern bathroom/shower-room? The following are the norm, although all may not be needed in the same location:

1 A light point/s with switches, usually cord operated
2 A shaver socket
3 A shower unit with or without a pump
4 An extract fan
5 A ceiling or wall mounted infra red or halogen heater
6 A heated towel rail
7 A Jacuzzi or whirlpool system

Many of these items are specifically designed to be installed in the various zones and in consequence, present no problem. The following comments illustrate what and where electrical systems/equipment/accessories may be installed in the various zones.

Zone 0
(in the water)

Wiring systems
Not easy to think of any *common* system that would be run in the water, however, only Zone 0 wiring should be installed in Zone 0.

IP codes for equipment
Only IP X7 equipment (immersion in water) permitted.

Switchgear and controlgear
Surprise! Surprise! None permitted unless incorporated in equipment designed to be in this zone.

Fixed current using equipment
Only that which is specifically designed to be in this zone.

Zone 1

(above zone 0 to 2.25 m and below zone 0 if area is accessible, e.g. if the bath panel is removable *without* the use of a key or tool, otherwise, this is outside all the zones)

Wiring systems

Only wiring for this zone or zone 0.

IP codes for equipment

IP X4 (splash proof) or
IP X5 where cleaning is carried out by water jets.

Switchgear and controlgear

Apart from items specifically designed for this zone, only SELV switches and the insulating cords of pull cord switches (not the actual switches).

Fixed current using equipment

Water heater ⎱
Shower pump ⎰ This is surely a shower unit (no RCD required).
Other suitably designed items provided the supply to them is 30 mA RCD protected.
SELV equipment.

Zone 2

(600 mm from edge of zone 1, & above zone 1 to 3.0 m)

Wiring systems

Only that appropriate for equipment in this zone & zones 0 & 1.

IP codes for equipment

IP X4 (splash proof) or
IP X5 where cleaning is carried out by water jets.

Switchgear and controlgear

Apart from items specifically designed for this zone, only SELV switches, sockets and the insulating cords of pull cord switches (not the actual switches).

Shaver units.

Fixed current using equipment

Anything which is specifically designed to be in this zone.

Water heater ⎫
Shower pump ⎬ This is surely a shower unit (no RCD required).

Luminaire

Fan

Heating appliance

Whirlpool unit

SELV equipment.

Zone 3

(2.4 m from edge of zone 2 & above zone 2 to 3.0 m)

Wiring systems

Any (suitable for the zones)

IP codes for equipment

None

Switchgear and controlgear

Sockets – SELV only

Switchgear – any, e.g. Fused connection units, plate or pull cord, switches etc.

Shaver units.

Fixed current using equipment
> Any
> If not fixed, e.g. Washing machine etc., should be fed via. A fused connection unit and the circuit protected by a 30 mA RCD
> No provision for portable equipment except SELV.

Outside the zones
(but still in the location)

Wiring Systems
Any

IP codes for equipment
Normal, i.e. IP 2X and IP XXB

Switchgear and controlgear
> Any
> Shaver units
> Only SELV sockets.*

Fixed current using equipment
Any

Supplementary equipotential bonding
This is a much misunderstood and misinterpreted subject and needs some clarification.

Clearly a bathroom, or shower room environment is hazardous due to body resistance being lowered as a result of water and, in

---

* Exception: When a room is not a bath/shower room, e.g. A bedroom which houses a shower cubicle, a normal socket outlet is permitted if it is protected by a 30 mA RCD, and is outside all the zones

consequence, it is important that dangerous potentials between exposed and extraneous conductive parts are not present in such areas.

In Zones 1, 2 & 3, the protective conductor *terminal* of each *circuit* supplying Class I and/or Class II equipment needs to be connected together with a *supplementary bonding conductor.* Bonding is not necessary between individual items on the *same* circuit, as the existing cpc acts as the bonding.

So, the earth terminal of a shower unit, either at the unit or its isolation switch would need a bond to an earth terminal on the lighting circuit.

Also within these zones, all extraneous conductive parts must be bonded together. Such parts would normally include hot and cold taps, metal waste and soil pipes (quite rare now), central heating radiators, exposed structural steelwork in leisure centres etc. Metal baths and shower basins, although quoted in BS 7671 Part 6, are not strictly extraneous as defined. Such items are unlikely to *introduce* a potential into any of the zones. As a general adage, if it is metal and inside any of the zones and not connected to any potential (earth included) from outside the zones, then no bonding is required.

Much of the extraneous bonding is already done by metallic pipe-work, provided it is mechanically sound and electrically continuous. Using a metal pipe as a bonding conductor where compression or plastic 'speed fix' fittings have been used!! may require extra bonding straps across these joints.

To complete the bonding there has to be a connection between the exposed and the extraneous parts. This is often achieved with one conductor from, say, the shaver unit to the basin taps. If all the circuits are bonded and all the pipes etc. are bonded then it only needs one bond from one system to the other to create the desired equipotential zone. So in Figure 48 the exposed conductive part of the towel rail is bonded to the other electrical circuits via metal pipes and other bonding. The join between the *dots* on this figure show the bonding requirements and illustrates just how little bonding needs to be carried out.

Where plastic pipework is used, even if the final joints to taps etc. are metal to plastic, there is no requirement to carry out any

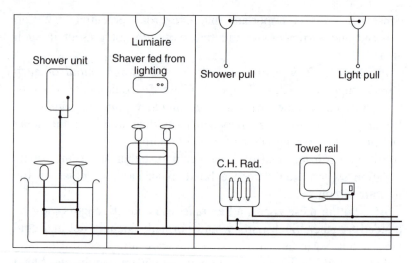

Figure 48 Supplementary bonding

bonding between extraneous conductive parts or between these and exposed conductive parts.

The electrical circuits still need the bonding between the circuits and to exposed structural steelwork etc.

Supplementary bonding conductors should be 2.5 mm^2 if they are sheathed or otherwise protected against mechanical damage (e.g. enclosed in conduit or trunking), or 4.00 mm^2 if not. Generally, contractors use 4.00 mm^2.

Supplementary bonding may be carried out close to the location, e.g. in an airing cupboard just outside the area or in a convenient place in a loft etc. Remember all screwed terminal earth bonding connections must be accessible.

Underfloor heating may be installed below any zone and must be covered with an earthed metal grid or sheath which in turn must be connected to the supplementary bonding.

Swimming pools

In a similar fashion to Bathrooms and Shower rooms etc. swimming pools are also divided into zones. In this case, however, they are zones A, B & C

Zone A is in the pool
Zone B extends 2.0 m horizontally from the rim of zone A and 2.5 m vertically above it regardless of the pool being above or below ground level. If there are diving boards, shutes or flumes etc. the 2.5 m height extends from their top surface.
Zone C extends a further 1.5 m horizontally from the edge of zone B and 2.5 m above ground level.

Now, what can we install in these zones?

Zones A & B

Wiring systems

Only systems supplying equipment in these zones are permitted.
Surface wiring systems must not employ metal enclosures or exposed metal cable sheaths or bare earthing or bonding conductors.

Switchgear and controlgear
None, unless the pool dimensions do not include a zone C. In these circumstances socket outlets are permitted if they are:

(a) BS EN 60309-2 (industrial type)
(b) More than 1.25 m from the edge of zone A
(c) More than 300 mm above floor level
(d) Protected by a 30 mA RCD or Electrical Separation.

Equipment
Only that which is designed for these locations.

Zone C

Switchgear and controlgear, and equipment
Socket outlets to BS EN 60309-2, switches, equipment and accessories provided they are protected:

(a) individually by Electrical Separation, or
(b) by SELV, or
(c) by a 30 mA RCD.
Shaver sockets.

IP Rating of enclosures

Zone A	IP X8 (submersion)
Zone B	IP X4 (splashproof) or
	IP X5 (where water jets are used for cleaning)
Zone C	IP X2 (drip proof) indoor pools
	IP X4 (splashproof) outdoor pools
	IP X5 (where water jets are used for cleaning).

Supplementary bonding

All exposed and extraneous conductive parts in zones A, B & C must have local supplementary bonding. This should also be connected to any metal grid covering underfloor heating.

Hot air saunas

Once again a zonal system has been used as per Figure 49. In this case the zones are based on temperature.

Wiring systems

It is unlikely that wiring other than flexible cords for the sauna equipment will be installed inside the sauna. These cords must be 180°C Thermosetting rubber.

Equipment

All should be at least IP 24

Zone A	only the sauna equipment
Zone B	no restriction regarding temperature resistance
Zone C	must be suitable for 125°C
Zone D	luminaries suitable for 125°C

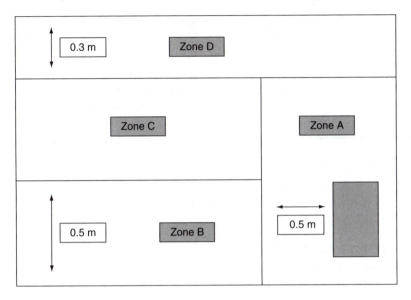

Figure 49

Switchgear, Controlgear and Accessories

Only that incorporated in the sauna unit or thermostats or thermal cut-outs.

Construction sites

Not as complicated as one may think. The only areas that require special consideration are where construction work is being carried out, not site huts etc.

So, let us keep all this as simple as possible. Clearly, construction sites are hazardous areas and in consequence the shock risk is greater. As we already know, 230 V socket outlet circuits need to disconnect in 0.4 s and fixed equipment circuits in 5 s.

This is alright for the site huts etc. but out on the site the only socket outlets permitted will be supplied from a 110 V centre tapped to earth single or three phase transformer and the disconnection time for such a reduced low voltage system is 5 s. Alternatively SELV could be used for hand lamps etc.

The only 230 V equipment permitted is fixed floodlighting and the disconnection time is 0.2 s. For fixed or moveable plant at 400 V three phase (cranes and lifts etc.) the voltage to earth (U_0) is the same as 230 V single phase, so the disconnection time is still 0.2 s.

We have already seen earlier in this book that touch voltage should not exceed 50 V. For construction sites this value is reduced to 25 V.

Wiring systems
Underground or overhead depending on the site, but whichever is used it must be protected against mechanical damage.

Isolation and switching
The main distribution board must provide all the normal facilities for isolation, protection against overcurrent, emergency switching etc.

Plugs and sockets/cable couplers
All should be to BS EN 60309-2.

Agricultural and horticultural premises
As with construction sites, not over complicated and much is very similar.

The requirements apply only to the areas outside the main farm-house, where the environment is hazardous and where disconnection times and touch voltages need to be reduced (livestock is killed at more than 25 V AC).

Wiring systems
Any, just as long as it is suitable for the environment, the minimum degree of protection being IP 44.

A high impact pvc conduit/trunking system would be appropriate in many cases as it is not affected by corrosion, is rodent proof and has no exposed conductive parts. However the system would be designed to suit the particular environmental conditions.

Switchgear and controlgear

Whatever!! As long as it is suitable for the conditions and that emergency switching is placed in a position inaccessible to livestock and can be accessed in the event of livestock panic. (Stampede!!!)

Protection against shock, fire and thermal effects

The standard method of EEBADS prevails, but with reduced disconnection times.

Protection against fire may be provided by the use of an RCD rated no greater than 500 mA, provided essential welfare of livestock would not be compromised by its operation.

It is also important that any heating equipment is kept clear of livestock and any combustible material and a distance of no less than 500 mm is recommended.

Electric fence controllers

Where mains operated controllers are used, they must be installed to ensure the following:

1. They should not be subject to adverse induced voltages from overhead power lines, or fixed to power or telecoms poles unless the pole carries the supply to the controller and the lines are insulated.
2. There should be no risk of mechanical damage or un-authorised interference and only one controller should be associated with any fence. Any earth electrode should be separate from any other system.

Restrictive conductive locations

These are very rare locations which could include metal tanks, boilers, ventilation ducts etc., where access is required for maintenance, repair or inspection.

Bodily movement will be severely restricted and in consequence such areas are extremely dangerous.

This section deals with the installation inside the location and the requirements for bringing in accessories/equipment from outside.

For fixed equipment in the location, one of the following methods of protection shall be used:

(a) SELV, or
(b) EEBADS but with additional supplementary bonding, or
(c) Electrical separation, or
(d) The use of Class II equipment backed up by a 30 mA RCD.

For hand held lamps and tools the socket must be supplied from a SELV circuit with the source of the supply outside the location.

Installations with high protective conductor currents

Generally, these installations are those involving IT and data processing equipment etc. or any system that may produce currents, all-be-it very small, that flow in *protective conductors* in normal service. These are not fault currents.

The size of *protective conductors* are determined in order to reduce the risk of shock and to avoid damage to conductors and insulation by excessive fault currents.

In consequence, *protective conductors* are quite capable of carrying the small currents produced by IT etc. equipment.

It is the *integrity* of the *connection* of these conductors that is important not the size. However, size can have an influence, in that a large conductor is more likely to be mechanically and electrically sound than a very small one!

Section 607 deals with equipment and circuits that may produce *protective conductor currents* in excess of either 3.5 mA or 10 mA.

In this section, there is a clear distinction between:

1. individual items
2. distribution and final circuits (that do not supply socket outlets)
3. final circuits feeding socket outlets. (Typically BS1363 13A socket outlets.)

The following figures illustrate the Regulations that apply.

For items of equipment that may produce a *protective conductor current* of more than 3 mA but less than 10 mA, the following requirements apply:

For items of *equipment* that may produce a *protective conductor current* of more than 10 mA, the following requirements apply:

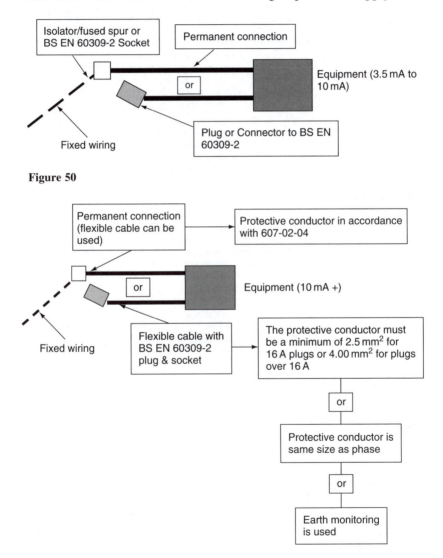

Figure 50

Figure 51

Distribution and *final circuits* where *protective conductor current* exceeds 10 mA (not socket outlet circuits)

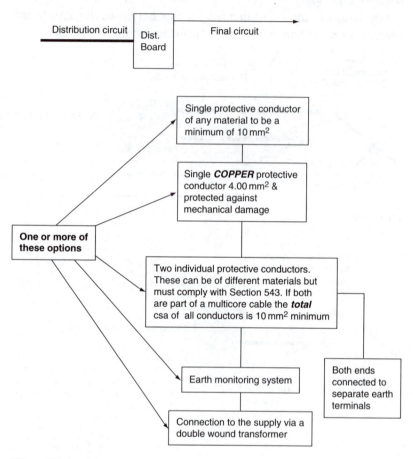

Figure 52

Ring and radial final circuits

For these circuits the existing CPC is of adequate size and all that needs to be ensured is that each outlet has TWO earthing terminals and the CPC's are connected one into each. Also at the distribution board the CPC's must be connected into different terminals. Spurs from a ring must be supplied with an extra CPC, and radial circuits must also have a secondary CPC wired in the form of a ring.

Residual current devices
Any RCD used must not operate due to collective protective conductor currents, so great care needs to be exercised in the selection, if any, of such devices.

Caravans and caravan parks

Caravans
These are the little homes that people tow behind their cars, not those that tend to be located on a fixed site. It would be unusual for the general Electrical Contractor to wire new, or even rewire old units. How many of us ever rewire our cars? In consequence, only the very basic requirements are considered here.

Protection
These unit are small houses on wheels and subject to the basic requirements of protection against shock, and overcurrent. Where EEBADS is used this must be backed up by a 30 mA RCD.

Wiring systems
The wiring systems should take into account the fact that the structure of the unit is subject to flexible/mechanical stresses and therefore, our common flat twin and three core cables should not be used.

Inlets
Unless the caravan demand exceeds 16 A, then the inlet should conform to the following, it should be:

(a) to BS EN 60309-2 (key position 6 h. This means the keyway on the socket and plug are at 6 o'clock!)
(b) no more than 1.8 m above ground level

(c) readily accessible and in a suitable enclosure outside the caravan.
(d) identified by a notice that details the nominal voltage, frequency and rated current of the unit.

Also, inside the caravan, there should be an isolating switch and a notice detailing the instructions for the connection and disconnection of the Electricity Supply and the period of time between Inspection and Testing (3 years).

General
Accessories and luminaries should be arranged such that no damage can occur due to movement etc.

There should be no compatibility between sockets of Low and Extra Low voltage.

Any accessory exposed to moisture should be IP 55 rated (jet proof and dust proof).

Caravan parks
We drive in to a caravan park for our holiday and need to connect to a supply of electricity for all our usual needs. This is little different to a block of flats (each flat representing a caravan) with a main intake feeding the various units.

Wiring systems

(a) preferably underground cable suitably protected against mechanical damage, tent pegs, steel spikes etc.
(b) if overhead then, 2 m outside the park boundary and 6 m above ground where there is vehicle movement and 3.5 m elsewhere

Switchgear

(a) supply equipment should be adjacent to, or within 20 m of the pitch.
(b) Socket outlets should be: to BS EN 60309-2; IP X4 (splash proof); at between 0.8 m and 1.5 m above ground; rated not less than 16 A.

(c) Each socket should be individually protected by an overcurrent device and a 30 mA RCD (although groups of no more than 3 could be RCD protected)
(d) If the supply is TNC–S (PME) the protective conductor of each socket needs to be connected to and earth rod.

Highway power supplies and street furniture

Lamp posts, illuminated traffic signs etc. Street furniture are not tables and chairs outside Bistros!!

These, like caravans, are mini-houses. Each unit has an intake position, overcurrent and shock protection, based on standard installations.

Whilst BS 7671 details heights of doors etc. above ground level, it is unlikely that any installer will need to be concerned, as the manufacturers will have dealt with these issues.

All equipment should be to IP33, and disconnection times are 5 s.

Locations with an increased risk of fire (BS 7671, Chapter 48)

These are typically areas:

(i) where combustible materials are stored or manufactured or are the by-product of a process, e.g. woodworking or textile factories, paper mills etc.
(ii) where the building construction itself or part of it is of a combustible material.

The requirements for such areas are based on preventing dust, fibres etc. from accumulating in enclosures and kept away from high temperatures. The following are the general salient points:

1. Enclosures should be to at least IP 5X (dust proof).
2. Enclosures should not attain temperatures above 90 °C under normal conditions and, 110 °C under fault conditions.
3. Surface wiring systems/cables should meet the Flame propagation requirements of BS EN 50265 and BS 4006-3.

4. Other than Mineral Insulated cable and Bus-bar trunking, all wiring in TN, and TN systems must be protected by a 300 mA, or less, RCD.
5. Heating devices and luminaries should be selected and erected in order to prevent ignition of combustible materials.
6. Cables, conduit, equipment, boxes distribution boards etc. should be selected and erected to avoid combustion of the building fabric (wooden walls, thatched roofs etc.).

THE 2004 VERSION OF BS 7671:2001 (INCORPORATING: AMENDMENTS 1:2002 & 2:2004) — THE CHANGES

1 The Electricity Supply Regulations 1988 have now been replaced by 'The Electricity Safety Quality and Continuity Regulations 2002'.

2 New conductor/cable colours (this appendix deals only with ac circuits).

From the 1st April 2004 new colours were introduced. Old *or* new colours may be used for new installations until 1st April 2006 after which only the *new* colours are acceptable. CPC colours remain the same.

These new colours are as follows:

SINGLE PHASE:

     Phase: Brown
     Neutral: Blue

THREE PHASE:

Phase: Brown or Black or Grey (if all brown or all black or all grey are used, they must still be identified brown, black & grey at terminations)
Numbers and letters may also be used to identify phases using L1, L2 and L3
Neutral: Blue

For a colour illustration of the interface between old and new please see the inside back cover.

ANSWERS TO PROBLEMS

Chapter 2

1 $0.152\,\Omega$
2 $18\,\text{m}$
3 $0.56\,\Omega$
4 No
5 Yes. Change the cooker control unit to one without a socket, thus converting the circuit to fixed equipment.

Chapter 5

For the factory design problem, the values obtained are as follows:

$I_b = 19\,\text{A};$
$I_n = 20\,\text{A};$
$C_f = 0.725;$
$C_a = 0.94;$
$C_g = 0.8;$
$I_t = 36.6\,\text{A};$
cable size $= 6.0\,\text{mm}^2;$
CPC size $= 2.5\,\text{mm}^2;$
$Z_s = 1\,\Omega;$
$I = 240\,\text{A};$
$t = 1\,\text{s};$
$k = 115;$
conduit size $= 32\,\text{mm}.$

INDEX

Index